NUCLEAR DETERRENCE AND DIPLOMACY

NUCLEAR DETERRENCE AND DIPLOMACY

Air Commodore
Jasjit Singh
AVSM VrC VM IAF (Retd)

and

Dr. Manpreet Sethi

Knowledge World

in association with

Centre for Air Power Studies

The Centre for Air Power Studies is an independent, non-profit, academic reseach institution established in 2002 under a registered trust to undertake and promote policy-related research, study and discussion on the trends and developments in air power, and related aspects of military power and strategy. Its publications seek to expand and deepen the understanding of air power and aerospace issues without necessarily reflecting the views of any institution or individuals except those of the author.

Jasjit Singh
Director
Centre for Air Power Studies
No. 1, Vasant Enclave
New Delhi-110057

Tele: (91-11) 26144100
E-mail: airpowerindia@yahoo.co.in

Knowledge World
5A/4A Ansari Road, Daryaganj, New Delhi 110002

Published in India by Knowledge World

355.0217

Nuclear Deterrence. Arms Control. NPT—India. FMCT CTBT—India. Ballistic Missile Defence. Nuclear Weapons India—Defence. Diplomacy. Jasjit Singh. Manpreet Sethi. Centre for Air Power Studies

ISBN: 81-87966-05-X

Typeset by Hardev Printers, Printed and bound in India by Bhavish Graphics. Published by Kalpana Shukla, Knowledge World, 5A/4A Ansari Road, Daryaganj, New Delhi-110002.

Contents

Acronyms

ABM	Anti-Ballistic Missile
ASAT	Anti-Satellite
AMM	Anti-Missile Missile
BARC	Bhabha Atomic Research Centre
BMD	Ballistic Missile Defence
BMDO	Ballistic Missile Defence Organisation
CBM	Confidence Building Measure
CBW	Chemical and Biological Weapons
CD	Conference on Disarmament
COSC	Chiefs of Staff Committee
CTBT	Comprehensive Test Ban Treaty
COPUOS	Committee on the Peaceful Uses of Outer Space
CWC	Chemical Warfare Convention
DEW	Directed Energy Weapons
ENDC	Eighteen-Nation Disarmament Committee
FMCT	Fissile Material Cut-off Treaty
GPS	Global Positioning System
HEU	Highly Enriched Uranium
IAEA	International Atomic Energy Agency
ICBM	Inter-Continental Ballistic Missile
ICJ	International Court of Justice
ICOC	International Code of Conduct
IDC	International Data Centre
IMS	International Monitoring System

INF	Intermediate Range Nuclear Forces
IRBM	Intermediate Range Ballistic Missile
JIC	Joint Intelligence Committee
JPC	Joint Planning Committee
KT	Kilo tonne
LEO	Low Earth Orbit
MAD	Mutually Assured Destruction
MIRV	Multiple Independently Targetted Reentry Vehicle
MTCR	Missile Technology Control Regime
NAC	New Agenda Coalition
NATO	North Atlantic Treaty Organisation
NFU	No-First-Use
NMD	National Missile Defence
NPR	Nuclear Posture Review
NPT	Non-Proliferation Treaty
NPTREC	NPT Review Conference
NWS	Nuclear Weapon State
NNWS	Non-Nuclear Weapon State
NSAB	National Security Advisory Board
NSC	National Security Council
NSG	Nuclear Suppliers Group
NWFW	Nuclear Weapon-Free World
OST	Outer Space Treaty
PAC	Patriot Advanced Capability
PAL	Permissive Action Links
PAROS	Prevention of Arms Race in Outer Space
PDD	Presidential Decision Directive
PTBT	Partial Test Ban Treaty
RV	Reentry Vehicle

SALT	Strategic Arms Limitation Talks
SBIRS	Space Based Infrared Satellites
SBW	Space Based Weapons
SDI	Strategic Defence Initiative
SLBM	Submarine Launched Ballistic Missile
SLV	Satellite Launch Vehicle
SORT	Strategic Offensive Reduction Treaty
SRP	Strengthened Review Process
SSMP	Stockpile Stewardship and Management Programme
SSOD	Special Session on Disarmament
START	Strategic Arms Reduction Treaty
THAAD	Theatre High Altitude Area Defence
TMD	Theatre Missile Defence
TNW	Tactical Nuclear Weapons
TVA	Tennessee Valley Authority
UNGA	United Nations General Assembly
UNSC	United Nations Security Council
WMD	Weapons of Mass Destruction
ZBM	Zero Ballistic Missiles

Preface

Nuclear diplomacy worldwide is in a state of flux. Nuclear deterrence is being overhauled to accommodate missile defence, and arms control is facing an unprecedented challenge. Treaties such as the Nuclear Non-Proliferation Treaty (NPT), Comprehensive Test Ban Treaty (CTBT), Strategic Arms Reductions Treaty (START), Anti-Ballistic Missile (ABM) Treaty, etc that had attempted, however imperfectly, to place certain restrictions on the horizontal or vertical development of nuclear weapons, are facing a crisis of legitimacy as the present status of each is very distant from what was intended. The contemporary situation is very different from what the world faced only about a decade ago, when standing at the threshold of the end of the Cold War, in 1991, one international analyst had caught the spirit of the changing times in these words, "These are the days of hope, not despair. It is time to strip away the complex and arcane strategic theory of the Cold War and start from scratch."

Barely fourteen years later, the optimism that was so tangible then has virtually dissipated to leave behind frustration and scepticism in its wake. Strategic theory no doubt is being stripped of the Cold War cloaks. But it is not certain that it is moving in any way towards a reduction of nuclear threats and risks. Clearly though, it is not headed, at least yet, toward a paradigm of cooperative security in a progressively less aligned world. Rather, the trend seems to favour unilateralism and an increasing salience of nuclear weapons, especially as a foreign policy tool, as well demonstrated by North Korea. There are also growing indications that most present day international eminent strategic experts are more grounded in so-called pragmatism, and accord far less credence to the goal of nuclear disarmament, although strategic logic still points to the military-strategic advantage of global abolition of nuclear weapons to be unambiguously in the interest of international security.

As a non-nuclear nation, India had a certain perspective on issues related to nuclear deterrence and arms control. It was not enamoured with nuclear deterrence since it was deemed integral to the existence and perpetuation of nuclear weapons, a reality that India abhorred and strove to change. Arms control was considered one tool to undertake this exercise, though India accepted arms control only as an interim measure, pending universal disarmament. Interestingly though, India had come to realise within a few decades of its independent existence that arms control measures, whether designed in a global context or those more directly focussed on India, almost inevitably under the rubric of non-proliferation, were meant to deny, restrain and limit its ability to move toward acquisition of nuclear weapons technology, material and arsenal. Indian nuclear diplomacy, then, worked on the basis of this perspective to negate the impact of these controls to the extent possible, and to further India's interests at various bilateral and multilateral forums. It tried to reduce the negative impact of arms control measures directed to deny it legitimate capabilities for its security, while searching for more durable, universally accepted solutions to the challenges posed by nuclear weapons.

However, on May 11, 1998, India altered its formal nuclear status. Naturally, this change in the "terms of reference" requires the country to cast a fresh look at both the operational and diplomatic dimensions of the nuclear issue. Engaged as New Delhi now is in the task of building a credible minimum deterrent, its standpoint on several issues related to nuclear deterrence and diplomacy is different from what it was in the past. The arms control and non-proliferation regimes impinge upon its security in certain novel ways and India's objectives vis-a-vis them need to be carefully assessed.

During the years since Pokhran II, several books by Indian authors have focussed particularly on the historical dimension of the slow march toward weaponisation. Some others have described the kind of nuclear strategy that India must follow for operationalising the deterrent. These have debated the intricacies of nuclear strategy, the structure of command and control that must be set up, the nuclear targetting philosophy that should be followed, and the kind of arsenal that must be built. However, most of them have either ignored or made only a passing mention of the stance that nuclear India must project on its deterrence strategy, arms control and disarmament in a coherent manner. This diplomatic projection, however, is critical since it plays an important role in enhancing the credibility of nuclear deterrence. In fact, given the fact that India conceives primarily a political role for its nuclear weapons, and that deterrence is a mind game, the importance of this dimension cannot be underestimated.

Therefore, apart from the operational issues, it is crucial to determine the "fix" for India in the contemporary nuclear maze and consider how it must react to the unfolding realities. How does it perceive nuclear deterrence and its operation in the regional context? As a state with nuclear weapons, but one that is not accepted as a nuclear weapon state under the NPT, and the associated non-proliferation regimes, how can and should India participate in arms control? In fact, what stakes should India have in the present non-proliferation and arms control regimes? Should it strive to retain the non-proliferation edifice as it exists, on the pillars of the NPT, CTBT, etc.? Or, could it manipulate its future form to meet national interests? What practical initiatives can India take to push forth the realisation of a nuclear weapon-free world? Or should it, as argued by some, abandon the goal of nuclear disarmament even if serves our national interests but may not be feasible in the foreseeable future? Can it strike a balance between *realpolitik* and idealism? And what does each of these mean for policy and our interests? These are some of the issues that require careful consideration if India is to play a meaningful role as a responsible nuclear actor.

The contemporary turmoil in the concept of nuclear deterrence and its consequent impact on arms control, disquieting as they are for international security and stability, nevertheless do offer some advantages to nuclear India. New Delhi can make use of the flexibility that the present state of nuclear flux offers to undertake two crucial tasks:

- Carefully assess and ascertain its own position on key nuclear issues. Given that the nuclear weapon states currently hold no clear and rigid positions on traditional non-proliferation and arms control, India is fortunately free of pressures to bind itself to any of the treaties such as the CTBT and FMCT (Fissile Material Cut-off Treaty) immediately.
- Seize the initiative to progressively mould the form of the nuclear structure to have it emerge, in the future, as a dispensation most suited to national interests.

The essays in this book attempt to examine the changing parameters of nuclear deterrence and existing non-proliferation and arms control structures in order to suggest positions for a nuclear India in the case of each. Nuclear weapons strategy cannot be de-linked from the use and utility of conventional military power; and, in turn, the conventional military must factor in the nuclear dimension wherever it impacts the operational environment. Very often we tend to study conventional war-fighting as something separate from nuclear deterrence. But the two are deeply intertwined and each would have a profound impact on the strategy and diplomacy conditioning the other.

At the same time, it is essential to understand the traditional Indian perceptions on nuclear deterrence and arms control and the degrees to which these may be undergoing changes since May 1998. Events of the recent past have raised many questions about the future of the NPT. What emerges from such an examination is that the efficacy of the treaty is in jeopardy not so much because of the countries that remain outside it (India, Pakistan, Cuba and Israel), but because those within it do not want to uphold their commitment to non-proliferation and disarmament. North Korea's decision to go ahead with its nuclear weapons and missile programmes, and its announcement of withdrawal from the NPT yet again, place a new strain on the treaty, and on non-proliferation in general. These stresses shall make themselves evident in the future Preparatory Committee meetings and in the 2005 NPT Review Conference. At the same time, one of the most recent global arms control measure, the CTBT, has been in a coma. In fact, with the US going ahead with designs of new categories of nuclear weapons and the possibility of resumption of nuclear testing in the future, the entry into force of the CTBT is a distant possibility. India would need to consider various implications of these developments within the larger phenomenon of the changing paradigm of arms control and strategic stability. This also requires a look at the FMCT that was expected to occupy centre-stage after the conclusion of the CTBT. However, it is nowhere near any final draft, and nor is it moving forward in any noticeable manner. Rather, it remains caught in a maze of difficult political and technical questions. We need to understand the issues that are currently exercising the minds of the negotiators and their implications for us.

Ballistic missile defences are becoming a reality but their implications for the future of strategic stability, nuclear deterrence and diplomacy are far from clear. Progress in missile defences is likely to push the world toward weaponisation of space and possibly a new arms race. Given the current trends in a world that is far more unpredictable and multipolar, can arms control be taken into outer space, to preempt or prevent such a possibility? If the fate of post-Cold War dissipation of arms control is any sign, the prospects are politically dim, in spite of the need being as urgent as it ever was. But as may be seen from US policies, there is less and less inclination to follow a path toward such a goal, and instead what we are witness to is increasing unilateralism that eschews bilateral or multilateral controls and seeks to retain maximum flexibility. In the process, certain kinds of nuclear dangers could well be on the rise, despite the actual numbers of warheads coming down. Reductions in the numbers of nuclear warheads can be useful only when they bring about simultaneous reductions in the threats, in the demands for Weapons of Mass Destruction (WMD), and in the sources of insecurity.

It is, therefore, necessary to take an incisive look at the reasons, not just moral but also practical and security related, that compel India to continue its pursuit of universal nuclear disarmament. This book historically traces the major Indian initiatives on nuclear disarmament. It examines the factors responsible for bringing about a change in mood on the subject that has pushed the country to find greater prudence in accepting nuclear deterrence in the short term. And to search for the type of deterrence that would suit our aims and objectives over the short and long terms. The conclusion is clear: the only true and durable basis for national and international security lies in a world free of nuclear weapons, an objective that India must continue to strive for.

Indian nuclear diplomacy is, therefore, tasked with the challenging responsibility of rationalising an operational nuclear deterrent with arms control en route to universal nuclear disarmament. All momentous changes in human history have come through changes in belief systems and the way people think about such issues. Nuclear weapons cannot be abolished unless people recognise that in an increasingly globalised world and interdependent society, the growing vulnerabilities of nations would require that we factor these into our thinking process. The steps to reach the destination of a world free of nuclear weapons would require us to rewrite the software programme of our belief systems through a gradual devaluation and delegitimisation of nuclear weapons so that current thinking and mindsets on nuclear weapons can gradually be eroded.

This book arrives at a historical juncture when international nuclear behaviour is no longer characterised by the luxury of predictability. Paradigm shifts in the operation of deterrence and arms control are certain to impact upon the state of non-proliferation and the future of disarmament. Evidently, the issues are only likely to become more complex, as countries work out their permutations and combinations of deterrence and defence against WMD. It is against this difficult backdrop that this book makes a sincere attempt to suggest Indian positions on nuclear issues related to deterrence and diplomacy.

As nuclear India finds its way in tomorrow's world of nuclear deterrence and arms control, and hones its nuclear diplomacy to push forth India's interests, several challenges lie ahead. Unfortunately, the dawn of a nuclear weapon-free world still eludes us. But we hope this book will contribute to strengthening the voices of sanity and help us to secure the future of our coming generations.

New Delhi
15th October, 2003

Jasjit Singh
Manpreet Sethi

1
Defence in a Period of Change

Jasjit Singh

In the context of our present inquiry, we can identify three benchmarks of changes of a fundamental nature due to nuclear weapons having come to play a crucial role in the defence environment: China's acquisition of nuclear weapons in October 1964, Pakistan's nuclearisation by end-1986 or so, and the May 1998 nuclear tests by India and Pakistan along with the declaration of being nuclear weapon states. While the change in the security environment due to Chinese nuclearisation perhaps occurred quickly and without adequate prior notice, the slow march of Pakistan toward acquisition of nuclear weapons had given enough notice of the change taking place. Both these changes no doubt were taken into account in shaping the policy regarding acquisition of nuclear weapons and the doctrine for them. But it is less clear whether any change in overall defence strategy to deal with the altered environment was ever initiated.

Change is the only constant in nature. A living organism becomes immune to change only when it dies. Change, therefore, in the field of national security and defence is something that must occupy our foremost attention so that requisite adjustments—changes—in strategy to deal with the implications of changes taking place are made. On the other hand, military organisations are historically noted for their conservatism besides the inevitable human inertia and unwillingness to change. This conservatism is understandable since few prudent people would like to take chances in such a key area of human enterprise as national security and defence. The risks of error in dealing with change can be momentous and the human tendency to stick with proven methods and techniques normally works against accepting change. International changes are the preferred route.

Changes also take a long time to be absorbed into military institutions and doctrines. Militaries are intrinsically conservative and traditional. This is understandable considering the risk of failure in deriving the correct assessment. The comfort level with existing systems is normally high in all human endeavours and change is not easy to assimilate. The nature of culture and institutions to identify change and manage it also plays a crucial role. On the other hand, military strategies are perpetually seeking to take advantage of the prevailing situation by resorting to changes in strategy and tactics, often building these on new technologies and capabilities. This is why, given the time, resources and perseverance, a military planner would almost inevitably find a way around the obstacles and weaknesses of any existing capabilities or circumstances. This phenomenon is most visible in the historical tussle between offence and defence, with the balance of advantage tilting in favour of one or the other depending upon circumstances, technology and strategy.

If change is a continuous and continuing process, then how should it be approached by defence strategy?

To begin with, and even at the risk of stating the very obvious, one must first identify the elements and essential features of change, the nature of change and its consequences for defence strategy, tactics and technology. And this must be time-sensitive. It would be impossible to craft a successful strategy to deal with changes taking place if such identification does not take place within a time horizon of relevance. It would naturally be too late to be useful if such recognition comes after change has impacted on the outcome of defence operations. Such identification would naturally have to be accompanied by adequate understanding of the changes involved and their implications for defence policy, at both the planning and operational levels. But the greatest challenge normally arises from the issue of accepting change and the need to make requisite changes in strategy. The "mindset" problem has affected military history more than the actual issues of technology or fighting capability. The mindsets of the colonial period or those of the Cold War persisting into the decades after change had taken place stand out in history.

Lessons of History

In our case, there is also another problem. This arises within the universal problem of assessing change which requires drawing the right lessons from past experiences and identifying the potential impact of change on future defence strategy. Lack of historical sense and history writing lead to gaps in understanding that have to be filled in by intuition rather than an objective study. The absence of institutional memory and detailed analyses worsens the problem of coping with change in general and many lessons have to re-learnt. For example, what lessons could we draw from the Sino-Indian War forty years ago which marked a watershed in the way we would need to fight a war in the high Himalayas?

One doesn't have to go far to look for examples. Suffice it to point out that Israel's war in Lebanon in Beqa'a Valley in 1982 clearly indicated that the technology, weapons systems and forces used in the operations had existed for quite some time and knowledge about their capabilities was available across the world. But what became unique was not only the way Israel used them, but the failure of a wide spectrum of military organisations worldwide, including those of Syria (which lost 86 combat aircraft in less than 48 hours), to absorb the nature of monumental changes taking place in warfare. Even more notable was the surprise (and sudden admiration) for the way the United States fought the Gulf War in 1990-91 although the essential change that took place between the Israeli operations in Beqa'a Valley and the Gulf War was one of scale rather than one of substance in the way war was fought. Closer home, it is not clear whether the factor of change in China's military strategy after it acquired nuclear weapons in 1964 was given any serious consideration by our defence planners. Or for that matter, what are the implications of China's ongoing military modernisation on our defence policy?

Even more important, it is not known how far the doctrine of "deep strike" of the late 1980s was consistent with the knowledge that Pakistan had acquired nuclear weapons by 1987 (and had even held out nuclear threats). Pakistan's acquisition of a nuclear deterrent was a material change in the military scenario of the subcontinent, perhaps even more than that of China in 1964, if for no other reason than the fact that its ideologically driven hostility of a permanent nature toward India is deep, intense and emotive, frequently erupting into aggressive violence and armed aggression across the borders. Pakistan's "low cost" option of a covert war through terrorism under the nuclear umbrella since 1983 in Punjab and since 1988 in Jammu & Kashmir is a clear example of a major change. It was obvious by the late 1980s that the type of war we would have to fight, or could fight successfully, had altered fundamentally. But it does not seem to have made any known impact on defence strategy and planning till at least the Kargil War.

A second dimension of managing change revolves around the issues of developing adequate tools and response, and applying them in practice. If we see the process of identification and *understanding* of the nature of change and its implications for defence strategy, the issue of *acceptance* of change, evolving appropriate *responses* for application in the short and long-term perspectives in totality, then it is clear that the process of defence strategy formulation is far more important than the attention that it seems to have received in our country so far. This concern is borne out by the fact that the institution that would normally be the lynchpin of such activity would be the strategic and defence planning apparatus. Its core would, of necessity, have to be based on another institution that would undertake assessment of the changes taking place or anticipated.

Identifying Change

It is in this context that we need to note the major changes that have been taking place in the arena affecting national defence. Firstly, the global and historical trend indicates a change in the very nature of warfare and armed conflict. Traditional wars for conquering territory are already confined to the archives of history simply because the rationale for such wars no longer exists, or does not exist in the manner it functioned earlier. Resource wars are extremely unlikely except perhaps in relation to energy resources.[1]

The biggest factor of change that we must deal with, even if only to draw appropriate lessons for managing change in its totality, is the nuclear factor and its impact on our defence strategy. Nuclear weapons have limited the aim, scope and extent of war among states that possess such capabilities (and their allies) because of the tremendously destructive potential of such weapons.[2] As Martin van Creveld writes, "From Central Europe to Kashmir, and from the Middle East to Korea, nuclear weapons are making it impossible for large sovereign territorial units, or states, to fight each other in earnest without running the risk of mutual suicide."[3] Total or even full-scale and unlimited conventional war has been relegated to the backyard of history because the age of imperial colonies, which was a major factor in making war a global phenomenon, has long passed away. States simply do not have the means to conduct a total war, unless they use nuclear weapons. And this would result in mutual assured destruction, nullifying any possible political goal for which it might have been started.

But modern states are also increasingly vulnerable even to conventional war. Conventional weapon attacks on nuclear power stations, for example, could result in a hundred Chernobyls in Europe. Modern precision guided weapons make it possible to execute damage of the scale carried out during the strategic bombing campaign in World War II with a fraction of the earlier air effort.[4] It is not merely the highly organised industrially developed states that would be so vulnerable to conventional warfare. The developing states are even more vulnerable because of the relatively few high value assets they possess which have been acquired through investment of scarce resources. The case of Iraq, where development has been set back perhaps more than three decades, first by Iranian air attacks, later by the US-led coalition air campaign and now the recent war waged by the US-UK coalition, is a prime example.

If the proposition that nuclear war and total global war are no longer viable as an extension of politics by other means, the only choice available to states to use destructive force for political purposes is through one (or a combination) of the following:

- *Limited Conventional War,* which could be full-scale (in the sense of using all elements of military power) but restricted in time and scope, or limited in the quantum and pace of application of firepower, possibly spread out in time, in what could possibly be termed as a war in "slow motion!" Political and military goals would be limited but synergised in a way that optimise them in our favour by maximising the strengths and minimising the effect of the weaknesses. No "decisive" military victory (in classical terms) would be possible against a state armed with nuclear weapons without seriously risking an escalation to the level of a nuclear exchange.
- *Sub-Conventional War* with military type weapons (including low-intensity war and war through terrorism), of the type that Pakistan has prosecuted during the past 20 years, below the nuclear threshold, and at a level of pain that does not call for a major conventional military response, leave alone an escalation to nuclear weapons use. There is an obvious risk that if the level and intensity

of sub-conventional war (as much as that in the case of conventional war) rises rapidly in time and space (as happened in the case of September 11, 2001, in the United States, and December 13, 2001, in India) escalation to conventional war would be probable.

- The use of *coercive military force* without necessarily leading to war, of which the military mobilisation and the threat of war by India after September 13, 2001, was but one example. This would have its own limitations and effects. It needs noting that historical evidence indicates that even successful application of coercive military power normally produces only short-terms gains without necessarily leading to a permanent change in the adversary's behaviour pattern.

In all these cases, care would have to be taken to ensure that the risks of escalation are contained. This in itself would tend to limit the scope of force being applied. What, however, needs to be remembered is that escalatory tendencies would increase if force is applied rapidly, which could cause the psychological pressures on the enemy to escalate. On the other hand, force applied in a graduated manner may not achieve the desired results within a desirable period. The overall result has been to shift the paradigm of war downward from nuclear and total war to conventional and sub-conventional wars with severe limitations on how such wars could be conducted if escalation to the nuclear levels was to be avoided. The key here is to ensure what has come to be known as "escalation dominance" throughout the spectrum of conflict which should also ensure that initiative to apply military power in time, space and scope would be retained.

Defence Through Deterrence

The strategic priority and core security interest of India will continue to be the prevention of war. As it is, the probability of a regular full-scale war in the foreseeable future is low. But if it does take place, its impact is likely to be far greater than that of earlier wars. It also needs to be noted that the next war, if and when it takes place, is most likely to be a limited war (essentially what the Chinese military has been referring to as a "local border war" since the mid-1980s) because of the nuclear factor operating in relation to both our major neighbours. Deterrence, therefore, would be the central pillar of defence strategy and would play a major role in preventing war as well as in fighting a limited war.

If Deterrence Fails?

The credibility of deterrence actually depends upon the credibility of a viable strategy which would lead to success in case of deterrence failure. Thus, defence strategy must address the issue of war aims in political terms and draw from it the military aims and objectives.

Three major elements of national power and capability can be identified which would constitute the centre of gravity of the adversary's national power. Damage and destruction of this centre of gravity would lead to severe degradation, if not total collapse, of the ability of the enemy to continue on the path of policy that necessitated the war as a response in the first place. These may be summed up as:

- *Territory,* whose loss would force the enemy state to shrivel and collapse as a sovereign state. A number of political and military factors have reduced the salience of territory as the centre of gravity of the enemy's power base. The risk of escalation to the nuclear level, however, would impose a major limitation on occupying the enemy's territory. Even where the target country does not possess weapons of mass destruction, the costs of keeping politically aware populations under alien military control would far outweigh the advantages of occupying territory by force. The challenges faced by the Anglo-American coalition after winning the war in Iraq are symptomatic. Territory, therefore, as a military-political objective of war must take this into account. The importance of territory would also depend upon its importance to the enemy, and the nature of the enemy system. In the overall analysis, territory has not been eliminated from the political aims of war, although its importance as the centre of gravity has dissipated.
- *Military power,* whose destruction at a substantive scale would leave it incapable of defending itself and, hence, force it to accede to the demands of the adversary. The Clausewitzian concept of war visualised the destruction of the enemy's military power as a key factor in winning the war to impose your will on the enemy. Destruction of the military forces also opens up a country

to greater vulnerabilities. However, possession of nuclear weapons implies that a country would seriously consider the option of using them if major components of its military power are destroyed or about to be destroyed since this could adversely affect the very survival of the state. This factor has been a major incentive for acquisition of nuclear weapons for defence. In fact, countries would hesitate to use military force in spite of even overwhelming military superiority where the target country possesses even rudimentary nuclear weapons capability. The case of North Korea, which is estimated to possess a couple of weapons as compared with that of Iraq with no weapons, is obvious. This limitation, therefore, drastically reduces, if not completely eliminates, the probability of a decisive military victory even as a rational goal, leave alone its achievement, if escalation to and beyond the nuclear threshold is to be avoided.

- *Social and economic assets,* attrition of which could impose a heavy burden on the adversary's capability to continue with its policies. Damage and destruction of leadership, social system, economic capabilities and other national assets would impose political, economic and social costs on the enemy state that would have to be factored into its policy choices in continuing along the policy path that justified the war as an instrument of policy to begin with. This is the area that offers the greatest potential for application of military power in a calibrated and/or graduated manner in a way that the immediate punishment does not warrant escalation to the nuclear (or even full-scale conventional) level, but the cumulative impact would impose costs that necessitate changes in policy. Developing countries are significantly vulnerable to strikes on economic infrastructure like power stations, key bridges (especially in mountains), major economic installations and other high value targets, the destruction of which would impose high economic and psychological costs on the enemy, particularly if he were unable to respond in kind. However, care would have to be taken to ensure that such destruction is not concentrated in time since this would tend to push the enemy to overreact. By the same logic, some of the impact of such military action would reduce when it is spaced out in time. But a clearly noticeable attrition campaign would lead to costs being raised even to levels which may be seen as very high.

Deterrence Limits and Capabilities

Deterrence may be achieved through *denial* or *punishment* (or a combination of both). All components of military power are capable of providing deterrence through denial, especially if close coordination among the three Services is achieved. However, it is in the punishment dimension that the individual components of military power start to have different capabilities and roles.

Land forces have been in the past, and will continue in future to be, the core of deterrence through denial. But they have very little capability to provide deterrence through punishment except through defeat and destruction of the adversary's land forces before any other punishment can be visited on the enemy. But in a limited war, such a situation may not be possible without the risk of escalation across the nuclear threshold. A "decisive victory" which would produce the conditions for imposing punishment is no longer a realistic scenario in the presence of nuclear weapons. This also, therefore, raises the question of escalation control which is so critical in a limited war between countries possessing nuclear weapons. Once engaged in combat, land forces cannot be disengaged unless one side or the other concedes defeat, or a ceasefire is agreed upon. Consequently, escalation control is poor where land forces are employed.

In our context, we also need to consider the scenario where land forces strike inside the territory of the adversary after deterrence has failed and war has been initiated. There would be a high risk of escalation to the nuclear threshold if a deeper strike takes place although the actual depth for that threshold would be a function of the importance of the area and the rapidity of advance where the intrusion was taking place. The threshold in a barren desert area would naturally be higher than, say, populated areas of Punjab. The issue that defence planners will have to come to terms with is the actual scenario where, if a nuclear threat is held out, it should be treated as credible. **The challenge for defence strategy is how to apply military power, achieve a level of military success and consequent political advantage, but keep the war below a rationally credible scenario of nuclear weapons use.**

Land war strategy itself necessitates that we closely examine the way war would be fought. The options and their implications could broadly be summarised as follows:

- Destruction of enemy military power through whatever strategy is adopted would lead to rapidly deepening the sense of vulnerability and despair which raises the probability of relying on nuclear weapons. This is the classical approach to military conflict but which in a nuclear scenario would also increase the risk of nuclear exchange. This is what Pakistani military leaders term as the "force threshold" of nuclear weapons.
- Military thrust into enemy country which would lead to loss of territory would raise serious concerns, which, if the loss was of sufficiently valuable territory, would constitute the "space threshold" of nuclear weapons, as stated by the Pakistani Army leadership.
- Multiple thrusts across the border to comparatively shallower distances in preference to single (or a couple of) deeper strikes would also result in loss of territory, but may not be seen as acute enough to push the situation close to a nuclear threshold. Yet politically such loss would be a material factor in any post-conflict negotiations. This may require changes in force structures, doctrine and tactics.

On the other hand, naval forces would provide a balanced set of capabilities for deterrence, by denial as well as by punishment. Similarly, naval forces possess a significant capability in terms of escalation control. This attribute in fact had been invaluable in "gun-boat diplomacy." The ability to engage and disengage at will had provided naval power the ability to select the place and time of benefit to it for waging war, and this was a major factor in ensuring that the most successful colonial powers were in fact maritime states. But in spite of increasing capabilities of land-attack from the sea, which are particularly neutralised by extension of sea denial capabilities by the adversary, naval forces are restricted to the maritime environment. This limits the scope of utility of naval power for deterrence through punishment. Every capital warship constitutes a major strategic asset, the loss of which would rapidly ratchet up pressures for use of higher levels of destructive force. Military technology has enhanced lethality and destructive capacity of naval power at ever increasing longer range. But technological developments like the anti-ship missiles, capable of being launched at long ranges from surface, sub-surface or aerial platforms, pose serious challenges to the survival of surface warships, especially if they are not protected by suitable means like integral air defence and anti-submarine warfare capabilities. The political-psychological impact of the loss of a capital warship would demand action, resulting in rapid escalation which may be difficult to control. The risk of escalation to the nuclear threshold, thus, would require serious consideration in the employment of naval power.

Air power, especially combat air power, intrinsically possesses attributes which provide it with powerful deterrence capabilities through denial as well as punishment. Aircraft can fly across national boundaries and geographical barriers to hit targets deep inside hostile territory. Loss of an aircraft to hostile action by itself would be unlikely to generate the type of pressures that the loss of a major warship would create. In some respects, the situation would be analogous to the loss of a main battle tank. The difference is that a single combat aircraft is far more capable of imposing damage and destruction at long ranges deep inside hostile territory as compared to a main battle tank which must directly defeat another land force before being in a position to impose any punishment beyond the attrition on the land forces in contact on the battlefield. Naval forces can achieve similar results with land-attack capabilities beyond the issue of direct confrontation. The impact of selective and discriminate destruction of carefully identified targets would increase the costs to the enemy. The imposition of such costs can be controlled and calibrated to manage the risks of escalation. The enemy's options would be to respond with aerial strikes or escalate beyond that to the employment of full military capabilities.

On the face of it this may represent a situation no different from a full-scale war. But two aspects deserve attention. First, the onus of escalation would be on the enemy. This is not a mere academic point. The enemy would be forced to calculate the costs and benefits of such escalation in relation to the continuation of its policies that triggered the need for punitive strikes in the first place. Second, the ability (as distinct from compulsion) to escalate would be a major factor in policy choice. The absence of such capability that provides a clear military advantage in any significant way would not allow such escalation any tangible advantage. To a large extent, the negative effects of such escalation can be contained by maintaining land and naval forces in a strong defensive posture while employing combat air power (even in possible conjunction with special forces across the borders) for strategic strikes. Highly calibrated escalation as well as disengagement control is possible with the use of air power. This would be a great asset in a limited war. In essence, this points toward the need to maintain a high level of conventional military capabilities, especially for roles and missions that serve the objective of cost-effective application of military power in a discrete manner to alter the policies of the enemy.

In many ways the war in Kashmir in the summer of 1999 offers a model for examination of the options for escalation dominance in real life scenarios. Pakistan followed up its covert war in Jammu & Kashmir by launching a limited war across the Line of Control in 1998-99 in the Kargil sector. Our use of air power was limited to the denial dimension only. The question, since the early 1980s, has been, how to undertake punitive application of military power which might change policies in Islamabad? Two cases may serve to highlight the core issues.

First, and hypothetically, what might have been our choices if the Indian Army, for good reasons and in spite of great feats of courage and sacrifice, had not been able to take the key heights in the Kargil War by end-June or early-July 1999? Our tactical situation would have been extremely difficult and embarrassing to say the least. The question here must be considered: what would be our options under these circumstances, and what would be the implications of those options? The following theoretical scenarios, among others, would have had to be considered:

- Escalate by counter-attacking in an area other than the Kargil sector. This was being advocated by many senior retired officers at that time. If followed, this course of action was most likely to have resulted in a full-scale war and heightened fears of a nuclear exchange, and a combination of pressure by the international community and domestic public opinion would have led in all probability to a ceasefire. This would have either resulted in the Line of Control being shifted eastward or Pakistan agreeing to withdraw across the Line of Control, claiming military victory and political concessions in return for which India would have been expected to concede some leeway on Kashmir in favour of Pakistan. Either way, the Kashmir issue would have been opened to UN/third party political intervention.
- Employ the naval forces to attack Pakistani land or naval assets in a controlled and calibrated manner. This would allow control

over engagement as well as disengagement. But it is not certain whether it would achieve commensurate results. There would have been every likelihood of Pakistan escalating to a full-scale war in response if it perceived it to be advantageous, or if it saw no alternative but to pursue that option. On the other hand, potential loss of a major warship by India would have had a disproportionate political-psychological impact on our defence strategy, forcing India to escalate in a dynamic situation where control and dominance could slip out of its hands.

- Use combat air power to attack sensitive targets across the Line of Control (for example, bridges across the rivers, power stations and other infrastructure in Pakistan Occupied Kashmir). This would indicate a resolve to escalate if the adversary was not willing to withdraw and India was forced to continue fighting from a position of disadvantage on land but with the ability to take advantage of the superior air power. There would still be a risk that the war would expand to full-scale level, but which could be kept limited by a strategically defensive "local border war." But it would be more likely that a calibrated air strike would extract an air power response only. Dominance, if not superiority, in the air then would be the key factor in deterring the use of air power by the adversary, or its ability to influence the outcome of the application of force. Long-term capacity building to achieve this aim would then become a critical necessity.

Management of Change

We are all managing change all the time. So the question is really about managing change in the context of national security in a way that produces results commensurate with the goals and objectives of national political-military strategy to serve our interests. There are three key elements in the management of change in defence strategy that must be given due weightage:

- Assessment of the nature of change and its implications for defence strategy.
- Planning for change, in force development as well as in force employment.
- Keeping the mind open to change.

Information, analysis and assessments of the nature of change and its implications on our interests and policies are the starting points of any effort to make change scientifically. This requires adequate intelligence inputs and continuing monitoring of the technological and operational environment that would impact on national security and defence. The roles and functions of information collection and intelligence assessment need to be undertaken by agencies and institutions separated from policy-making if objectivity is to be maintained.

The problem in our country is that the area of assessments for national security and defence perhaps remains the weakest in the area of inputs. This was demonstrated in the 1962, and later wars, including the Kargil War. In spite of a thorough look into national security management by Task Forces and by a specially constituted Group of Ministers at the ministerial level, we still do not have a national institution for carrying out net assessments. In the absence of such an institution, assessments related specifically to defence become limited even if they are undertaken in a comprehensive manner. Unfortunately, that is not so either. A Joint Intelligence Committee (JIC) was set up under the higher defence organisation in 1947. Unfortunately, the fiasco of 1962, instead of leading to reforms in the existing set-up, resulted in shifting the JIC out from being a committee under the Chiefs of Staff Committee (which itself was part of the governmental structure at that time) to being an entity under the Cabinet Secretariat. This might have proved advantageous if the JIC was structured as the overall national intelligence assessment body with specialist intelligence agencies feeding into it rather than being the sovereign prima donnas that they have continued to be. The net effect was that the JIC, with a staff of less than a dozen people with little or no background expertise in making assessment and mostly borrowed from other organisations, became a body for collation of intelligence information (as distinct from assessment) which the key intelligence agencies thought fit to share with it. It is to the credit of the JIC that it continued to deliver as much as it did in spite of the fundamental weaknesses so obviously prevailing.

But in reality the JIC simply dwindled in its capacity to undertake assessments for national security purposes largely because it was grossly understaffed, and lost out in bureaucratic leverage to the two main intelligence agencies, the Intelligence Bureau and later the Research and Analysis Wing of the Cabinet Secretariat. Both operated largely independent of each other. Over the years, the Research and Analysis Wing also diminished the staff and importance of intelligence and analysis of military significance. The Services intelligence organisations were totally dependent on inputs from the two intelligence agencies and collation of the same intelligence by the JIC, besides any information of tactical nature that they could themselves generate. Once again, a composite intelligence institution has been established under the Chiefs of Staff Committee, and one can only hope that it will undertake long and short-term net assessments of value to the making of defence strategy. This would require a large dedicated permanent staff for study and analysis, besides extensive studies from open sources to provide the backdrop.

The second aspect is that of planning to cater for change. This is required both at the level of force development and in the area of force employment. Planning for force development is crucial since it starts to define the capabilities at a future date. And existing capabilities in turn become the overriding factors in managing force employment. This becomes even more challenging if no flexibility is ensured. For example, our manpower employment policies today are such that 10-15 years would elapse if any meaningful changes were to be structured in tune with an altered defence strategy. The story is similar to that of intelligence assessments. A Joint Planning Committee (JPC) was set up as part of the new higher defence organisation in 1947 with "permanent staff, in permanent session" in 1947. However, the permanent staff was never established and there was no opportunity for the committee being in permanent session! The result was that we lost the experience of joint thinking and working at the defence planning level.

Another attempt was made by setting up the Defence Planning Staff of the Chiefs of Staff Committee in 1986. It did excellent work for some time. But it was grossly understaffed with no participation by expertise and manpower from non-military disciplines, and never given the importance it deserved, partly because by that time the defence Services were no longer integral to the government (and remained "attached offices"). The new reforms have now set up once again a joint planning staff under the title of Integrated Defence Staff. It remains to be seen how far this experiment will produce permanency and the requisite results. Interestingly, the Defence Planning Staff came up in 1986 at the time Pakistan was acquiring nuclear weapons and the defence scenario was changing fundamentally. But it is not clear what changes in defence strategy and policy were visualised in relation to this tectonic change.

This brings us to the third issue: how to deal with the problem of mindsets in managing change in defence strategy? This is a complex area and perhaps cannot be dealt with adequately in this essay. Suffice it to say that the **problem is both of culture and institutions.** This was identified by the Task Force set up to draw up the blueprint to set up the National Security Council (NSC) in 1998 and suitable recommendations made accordingly. In a way, subsequent events proved that this fundamental reality did not find favour with the decision-makers although the final NSC included two members (out of three) who had issued the Task Force report. The final decision appears to have been guided more by preservation of the *status quo* rather than with the creation of the culture and institutions to deal with change and future strategies. Subsequent actions in the government taking over the Indian Council of World Affairs (ICWA) as an institution of national importance and finally reshaping it into an extension of the Ministry of External Affairs further proved the unwillingness to develop institutions beyond the current issues and autonomous of the routine management of daily governance which has often boiled down to crisis management. If further proof of our unwillingness to develop institutions and the culture of empirical studies was needed, it was provided in the middle of the Kargil War. The NSC, in one of its rare meetings, decided in June 1998 to set up a dedicated think-tank to study Pakistan. Four years later, there is no sign of this taking off in spite of the proposal having received formal approval long ago.

What is required is continuous study; discussions and debate on what is changing, what are the implications, and what the defence strategy should be are the core of the process which could bring about the desired change. Unfortunately, reading habits seem to have died down and forward looking discussions and debate based on lessons drawn from the past and current military affairs are an exception rather than the rule. Few if any institutions exist beyond the official teaching task. Compare this with, say, the United States where even the youngest officers are enouraged to write and publish regularly. There are think-tanks, studying historical relevance as well looking at the future, in every teaching institution—often a number of think-tanks in a single teaching institution. For example, the National Defence University in Washington has eight think-tanks undertaking research, study and publications compared to the four teaching colleges in it. The picture in every Service is similar. China adopted a similar approach in the early 1980s when it started defence modernisation and recognised that change has to be managed at the culture as well as structural levels. We, however, remain sanguinely innocent of such measures to create the capacity to identify the nature of change and evolve options to deal with them.

Notes

1. Sanjay J. Singh, "War in the 21st Century," in Jasjit Singh, ed., *Air Power and Joint Operations* (New Delhi: Knowledge World, 2003).
2. Nine states (the USA, Russia, China, France, UK, India, Pakistan, Israel, and now North Korea) now possess nuclear weapons besides another 22 that are protected by nuclear weapons through formal treaty arrangements.
3. Martin van Creveld, *On Future War* (London: Brassey's 1991), p. 194.
4. See Jasjit Singh, *Air Power in Modern Warfare* (New Delhi: Lancer International, 1985).

2

Nuclear Deterrence and Arms Control

Manpreet Sethi

Tomes of literature already exist on the concept of nuclear deterrence and the rationale and philosophy that sustained it during the Cold War years. Superpower relations hinged on its precarious edge and many analysts believed that the world could avoid a third world war only because of nuclear deterrence. In the words of Walter Slocombe, the US under secretary for defence, "It is a remarkable fact that for almost half a century, the US and its allies faced the USSR...and did so armed to the greatest extent...and yet did not fight a large scale war...Can anyone really doubt that nuclear weapons had a role?"[1] Others too have credited nuclear weapons with the maintenance of global peace because on the many occasions that the superpowers did come close to a nuclear exchange, deterrence evoked the fear of mutual annihilation and set limits on their behaviour, cautioning them against taking certain kinds of actions.

The Changing Contours of Nuclear Deterrence

Perceived as an important component of strategic stability, attempts during the Cold War were made to maintain nuclear deterrence—whether through an arms race wherein nuclear countries tried to match their arsenals weapon for weapon; or, through arms control whereby countries sought to disarm each other, either equally or at least proportionately. Mutural Assured Reduction (MAD) and Strategic Arms Limitation Talks (SALT), and later START, were, hence, the respective manifestations of both the spectrums of nuclear deterrence between the US and USSR. The primacy of deterrence was, in fact, established over defence and institutionalised in the ABM Treaty. US strategic thinkers thought it prudent to put a stop to the development of anti-nuclear defensive technology because, ironically enough, greater nuclear safety was espied in the absolute and unchallenged capacity of a country to annihilate the other. The erection of defences, it was thought, would erode this vulnerability and thereby destabilise deterrence and lead to a build-up of offensive capabilities. Strange was the logic of security through nuclear deterrence, but the nuclear weapon states clung to it.

But, that was then, during the decades of the Cold War, when the adversary was clearly known and identifiable. In the present post-Cold War period though, the enemy is more diffused, less apparent and its *modus operandi* more flexible. These new threat asessments are believed to pose a challenge to the traditional forms of nuclear deterrence. Obviously, the huge stockpiles of nuclear weapons hold little deterrent value for meeting potential risks as arising from nuclear terrorism, use of nuclear weapons by rogue states or other non-state actors. Hence, primarily in the country of its origin, the US, the concept and operation of nuclear deterrence are being overhauled to adjust to the new threat perceptions. *The focus of nuclear diplomacy has shifted from conventional nuclear deterrence based on punishment or denial to one that is hinged on pre-emption, missile defence and the use of nuclear weapons to retaliate even against chemical or biological weapons.*

These changes in US nuclear policy, however, are not particularly reassuring for other nuclear and non-nuclear weapon states. Most countries, including US allies in Europe, perceive that American policies of preemption, deployment of missile defence and enlargement of the deterrent role of nuclear weapons would impact adversely on international strategic relations for four main reasons:

(a) The USA will no longer be deterred by the nuclear weapons of other countries. Mutual deterrence, therefore, shall become unworkable, leaving other countries more vulnerable and less secure. This would upset strategic stability, driving countries to buttress their security in other ways, possibly through acquisition of WMD.

(b) More secure under its protective missile shield, the US shall feel freer to unilaterally undertake preemptive strikes on others. Apart from the impact this would have on the credibility of the UN, a unilateralist approach to international relations will also undermine US commitment to multilateral arms control agreements. Some of this is already evident in the treatment being meted out to the ABM Treaty, CTBT, Biological Weapons

Protocol, Kyoto Protocol, etc. This unilateralism threatens the future standing of international arms control and non-proliferation regimes.

(c) The policy of preemption, even in the face of chemical and biological weapons, trashes the negative security assurances provided to non-nuclear states. Viewed against the overall context of US unilateralist and preemptive policies, this lack of assurance could drive other presently non-nuclear countries to reassess their commitment to not acquire nuclear weapons.

(d) The National Missile Defence (NMD) would allow the US the luxury to isolate itself from regional conflicts where it perceives no direct threat to its national interests. This thought is particularly disquieting for those countries that have long enjoyed the US' protective umbrella. It could compel them to consider security alternatives, including their own possession of WMD or submission to the erection of Theatre Missile Defence (TMD). Both of these would have repercussions for the security policies of other countries in the region.

The above-mentioned trends or developments impact on the nuclear relations, as they exist worldwide. Unfortunately, in this globalised world, it is not just the economies of the countries that are linked to one another, but also their security and military strategies are connected through "powerful ties of error and response."[2] Therefore, every action with nuclear overtones, whether emanating from a small or big country, has regional and global implications. For instance, the manner in which the US treated Iraq that was suspected of pursuing a nuclear weapons programme, and the manner in which the North Korean crisis is handled send out clearly demonstrative signals to others.

To illustrate the inter-linkages a little better, one may only have to logically think of the likely chain of actions that would unfold if a nuclear North Korea continues with its brinkmanship policies, reopens its plutonium reprocessing facilities and proceeds with more missile tests. Not only would Japan and South Korea consider this an immediate threat to their security, but the ripples would also be felt over a larger area. For one, Japan could decide to acquire a nuclear deterrent of its own. Or, as is even more likely, it would be pushed towards accepting TMD and a greater commitment from the US for its security. Moreover, the US would acquire an additional justification to continue its research and development of the NMD and TMD. This could possibly drive China to engage in a further increase in its offensive capabilities, including MIRVing of its missiles, and search for counter-measures against the missile defence systems. Chinese actions would cast their shadow on the nuclear policies of India, and even further on Pakistan, then on to Iran, and the rest of the Middle East. The string of consequences that would be unleashed, therefore, is likely to touch every aspect of global nuclear security.

Similarly, and as is already evident, US readjustment of its nuclear deterrence paradigm is certain to generate stresses for the entire nuclear edifice, including the arms control regime. The danger of proliferation of WMD to new powers is leading to the adoption of defences in the old ones. But defences would upset the balance of nuclear terror and stalemate arms control. This would, in turn, provide the impression that those in possession of nuclear weapons are going to hold on to them in perpetuity and breed further proliferation. This vicious cycle is difficult to break. During the Cold War, the existence of the nuclear arsenals was justified for the stability of nuclear deterrence. But, ten years after the demise of the Cold War, newer threats and logics have been found to recast nuclear deterrence as an important actor for strategic stability. The contemporary contours of US-laid nuclear deterrence, however, threaten to unleash a chain of events that could spin out of control and lead to a crisis in arms control.

The Crisis of Arms Control

After the close shave with the Cuban missile crisis, and once the superpowers had accepted the inevitability of living with the other's nuclear arsenal by the mid-1960s, negotiations sought to "stabilise" the nuclear stalemate. The USA and USSR, then, condescended to undertake arms control for four main reasons:

(a) To establish strategic stability and diminish the chances of use of nuclear weapons due to miscalculation.

(b) To reduce the adversary's capabilities without allowing one's own to be degraded.

(c) To showcase their restraint and reassure the international community of their intention to disarm some time in the future.

(d) To enforce non-proliferation.

Arms control was, therefore, undertaken once the adversary had caught up with a weapon and strategic stability was required to be established, or when a group of countries wanted to retain their monopoly over a weapon, or when advances in technology had made an earlier version of a weapon obsolete. Then, the countries sought to rationalise their arsenals through arms control. SALT, ABM, PTBT (Partial Test Ban Treaty), NPT, CTBT, etc have all taken shape to realise one or the other of these objectives. However, with the breakdown of bipolarity and the emergence of a new class of threats, the objectives of arms control have become less clear, at least in American strategic thinking.

Rather, US strategic circles are preoccupied with the following developments and are likely to consider future arms control through their prism:

- Dire threat assessments about the proliferation of WMD and their delivery capabilities.
- A perceptible sense of despondency with the non-proliferation regime. The US feels that despite all its efforts, proliferation is on the rise, and unstoppable.
- A decreasing faith in traditional diplomatic alignments and established multilateral fora to further the international consensus on non-proliferation.
- A greater propensity to use assertive military options through specific preemptive means to deter nuclear, biological and chemical attacks.
- An official determination to proceed with the deployment of the NMD, irrespective of the criticism from the international community.

Owing to these trends, arms control may not survive in its present form for very long. Instead, the current US predisposition appears inclined towards using approaches other than formal treaties for achieving arms control and strategic arms reductions. Some US officials have argued that given the reality of rapid advances in modern weapons technology; the evolving new threats posed by WMD; and the potential for dramatic and increasingly unpredictable international security scenarios, the US must preserve maximum flexibility in responding to circumstances. Therefore, they believe that it would be unwise for the US to undertake any new formal treaty obligations that would permanently limit or restrict this flexibility.[3] While this may serve the US interests well, the utility of arms control as an international confidence building measure and its binding effect on other countries could well be lacking.

Where does India find itself in this nuclear maze and how must it react to the unfolding realities? For a new nuclear country like India, that overtly established its nuclear status only five years ago and is still in the process of finalising its nuclear doctrine and strategy, the contemporary turmoil in the concept of nuclear deterrence and its consequent impact on arms control is particularly pertinent. As a country with nuclear weapons, but one that still aspires for a nuclear weapon-free world (NWFW) as the only true and durable basis for national and international security, Indian nuclear diplomacy is now tasked with the challenging responsibility of mating an operational nuclear deterrent with arms control, en route to universal nuclear disarmament.

India's Acceptance of Nuclear Deterrence: Ensuring Security in a Nuclearised World

Indian diplomacy, traditionally, was never enamoured of the concept of nuclear deterrence. If anything, Indian representatives at the UN consistently decried it for placing the international community on a nuclear precipice where nuclear war was always only a trigger away. Consequently, India has consistently argued that nuclear disarmament, rather than nuclear deterrence, must constitute the basis for lasting world peace and security. Its efforts in this direction spawned a number of initiatives some of which are listed below.

- As early as 1948, at the time of the consideration of the report of the newly created UN Atomic Energy Commission, India had proposed limiting the use of atomic energy to peaceful purposes and the elimination of nuclear weapons from national arsenals.
- In 1950, India used the UN to draw the attention of the world to the fact that the growing arms race was not only a threat to international peace and security but also a drain on human and economic resources that could be better employed in other developmental activities.

- India was the first country to call for a worldwide ban on nuclear testing in 1954.
- In 1956, India presented a memorandum to the Disarmament Commission suggesting certain concrete steps towards disarmament.
- In 1964, India took the initiative to place the item "Non-Proliferation of Nuclear Weapons" on the UN agenda and thereby sowed the first seeds of the NPT.
- At the UN Special Sessions on Disarmament in 1978 and 1982, India campaigned for outlawing the use of nuclear technology for military purposes and a treaty on the non-use of nuclear weapons and to make such use and threat of use a crime against humanity. It continues to regularly table the proposal at the UN, calling consistently for the commencement of talks on an international convention prohibiting the use and threat of use of nuclear weapons under any circumstances.
- In 1982, India proposed a permanent nuclear freeze to stop the further development and stockpiling of nuclear weapons.
- In 1984, at the height of the Cold War, India launched a Six-Nation, Five-Continent Peace Initiative to get the US and USSR back to the negotiating table.
- In 1986, India was among the strongest supporters of the Gorbachev proposal for the elimination of nuclear weapons.
- In 1988, then Prime Minister Rajiv Gandhi unveiled an Action Plan, a phased programme for the complete elimination of nuclear weapons within a specified time-frame.
- In 1996, India yet again proposed a programme of action for the elimination of nuclear weapons at the G-28 summit.

Evidently then, for nearly five decades, India continued to press on nuclear disarmament in every international forum. But the absence of any substantive progress on this front coupled with a deteriorating regional security environment made India go in for the acquisition of the nuclear deterrent. India had long upheld that for a country pursuing an independent foreign and security policy, potential challenges posed by nuclear weapons could be adequately addressed only by two means: global abolition of nuclear weapons; or, reliance on nuclear deterrence. In fact, since the time of Jawaharlal Nehru, independent India's first prime minister, the value of deterrence was recognised along with the horror of nuclear devastation. Every government, irrespective of the party, kept the nuclear option open, and steadily strengthened the technical capacity of deterrence, even though the actual production or deployment of nuclear weapons was eschewed.

In 1998, when India declared itself a state with nuclear weapons after the conduct of the five nuclear explosions, it was a conscious acceptance of the concept of nuclear deterrence. With Pakistan and China already nuclear capable, India found it necessary to demonstrate its nuclear capability in order to establish some sort of nuclear deterrence. Without going into the merits or demerits of this decision, it needs to be underlined that nuclear disarmament is still upheld as the long-term objective of nuclear diplomacy. India has still not abandoned its efforts towards total elimination of nuclear weapons through the establishment of a durable, non-discriminatory and comprehensive non-proliferation order. However, until such time as this order comes into being, India has predicated its security on nuclear deterrence that is now central to, and mutually supportive of, India's conventional military strategy.

However, the Indian conceptualisation of nuclear deterrence has certain unique discernible features, as highlighted in the Indian nuclear doctrine.

(a) *Minimum credible deterrence.* Acknowledging the role of nuclear weapons as more political than military, India defines their sole purpose as deterrents against the use and threat of use of nuclear weapons. Hence, a rejection of the concept of nuclear war-fighting frees India from the need to match the nuclear arsenal of its adversary weapon for weapon. Instead, India can afford to commit itself to a small force of survivable nuclear weapons that would deter an adversary from initiating military action in the first place. Official pronouncements have refused to be drawn into defining the minimum deterrent through any specific numbers. Rather, even the draft nuclear doctrine leaves the decision on the actual size and composition of the nuclear arsenal to the threat perceptions from the strategic environment, technological imperatives and the needs of national security.

(b) *Strategy of no-first use.* India's nuclear policy is underpinned by a clear commitment to "no-first-use" of nuclear weapons against nuclear-armed adversaries and the non-use of nuclear weapons against non-nuclear weapon states. Not being the first to undertake nuclear escalation, India does not need to keep its nuclear weapons in a deployed state of readiness for launch on warning. This drastically reduces the risk of use of nuclear weapons due to miscalculation or unauthorised use. At the same time, for deterrence to be credible, no-first-use demands the construction of a robust, and survivable deterrent capability that would restrain a first use by threatening a destructive retaliatory strike.

(c) *Assured massive retaliation.* A deterrence policy based on no-first-use depends heavily upon the unambiguous certainty that any first nuclear strike would lead to assured massive retaliation through a return punitive strike. For this, survivability of the arsenal is critical so that it can inflict damage unacceptable to the aggressor.

Indian nuclear deterrence, therefore, seeks to ensure the country's security and autonomy by providing the perception to the adversaries that India could absorb a first nuclear strike and yet be left with a viable number of survivable nuclear warheads that could destroy their major cities in a retaliatory strike. Thereby, it seeks to deter them from launching a first strike. The credibility of this deterrence, however, ironically enough, demands an operational nuclear force, a robust command and control, and a pre-determination of targets for a nuclear strike. In other words, a credible deterrent derives its trustworthiness from simulating a response to the likely breakdown of deterrence. It is this characteristic of deterrence that is expected to keep the peace, and yet none can deny the inherent dangers that underlie the stance.

For this reason, nuclear deterrence only in the short term, and nuclear disarmament in the long run must remain the unimpeachable objective of Indian security strategy. The predominance of realism in foreign policy-making is pushing Indian nuclear strategy towards the operationalisation of a credible nuclear deterrent. According to this logic, for the time being, and as long as nuclear weapons remain a factor in international relations, nuclear deterrence shall have to form a part of India's nuclear strategy. As the Indian prime minister said in Parliament on August 4, 1998, "India's nuclear tests were not intended for offence but for self-defence. In order to ensure that our independence and integrity are never jeopardised in future, we will have a policy of a minimum deterrent." But in the same breath he also exhorted our nuclear diplomacy not to lose sight of the objective of an NWFW. Nuclear disarmament has been seen by India as the only durable non-proliferation instrument and a key strategic and security goal. In a universally disarmed world, India's security lies along with that of the rest of the world.

Indian Views on Arms Control

However, in the interim, while India is required to strike a balance between nuclear deterrence and disarmament, how must it approach the issue of arms control? Engaged in the exercise of building a credible deterrent, should India eschew treaties such as the CTBT and FMCT that seek to restrain its nuclear options or should it continue to view arms control as a pragmatic means of eventually reaching nuclear disarmament?

Traditionally, India had been opposed to arms control measures on the grounds of principle since they seek to perpetuate the inequity of nuclear weapons with a few states. It also feared that the developed countries could use these measures to reduce, or at least freeze, its capabilities in a state of inferiority. The provisions of the NPT and its implementation, and subsequently, the manner in which the CTBT was sought to be imposed have done little to dispel this apprehension. Nevertheless, India has admitted a role for arms control as an interim measure until total disarmament can be reached. By enhancing confidence, arms control has been perceived as one way of achieving the eventual replacement of nuclear deterrence with nuclear disarmament. With this in view, India has always participated in the negotiation of all treaties on non-proliferation and supported other arms control measures between nuclear countries in the genuine hope of achieving the larger and more comprehensive objective of an NWFW.

This had been India's stance as a non-nuclear state. However, now as a state with nuclear weapons, how must India engage in the arms control process? Engage it must, since the very process would provide a tacit acceptance of its nuclear status, a reality that the NPT does not allow the other nuclear weapon states to acknowledge. The objectives of the Indian participation in the arms control process could be assessed against two benchmarks—the global and the regional.

At the global level, India's support for arms control would serve two purposes:

(a) To showcase its willingness to accept restraint, as long it is done in a non-discriminatory and verifiable manner.

(b) To genuinely use the process to arrive at universal nuclear disarmament through cooperative arms control that gradually builds confidence and transparency.

At the regional level, nuclear arms control presents a more complicated situation since the controls acceptable with China cannot be the same as those acceptable with Pakistan. In any case, if arms control, could be undertaken with the two nuclear adversaries of India, it would certainly meet the following objectives:

(a) To establish strategic stability.

(b) To serve as a regional confidence-building measure, especially if it can genuinely lead to the forsaking of a class of weapons that could reduce the risk of miscalculation and unauthorised use. Given the geographical proximity of countries in the region, a control on the delivery capabilities or tactical nuclear weapons would go a long way in enhancing security.

Conclusion

India has always maintained the need to establish an international order that can provide equal and legitimate security to all nations. Its initiatives towards the creation of an NWFW have been a part of this. However, in the absence of any hope to realise this state of affairs in a hurry, India was forced to accept the lesser evil of nuclear deterrence in order to safeguard its security in the face of two nuclear rival neighbours. However, Indian acceptance of nuclear deterrence does not in any way make the concept more endearing or less dangerous than what it was when India had not succumbed to its wily charms. The danger of maintaining nuclear arsenals and walking on the edge of the sword of nuclear deterrence are well known to India. At one level this calls for the operationalisation of the deterrent through the build-up of necessary armaments and delivery capabilities, as well as requisite command and control systems that raise the credibility of the deterrent. On the other hand, this credibility emanates equally from nuclear diplomacy since nuclear deterrence is as much a mind game that needs to indulge in the right posturing on issues of nuclear relevance.

At the same time, deterrence codifies and institutionalises a reactive cycle wherein each country responds to the possession of nuclear weapons with another through achieving the same possession for themselves. Underlying this philosophy is, therefore, the rationale for proliferation. In fact, in order for the stability of deterrence to be effective, it must be established through proliferation. The dangers of this can only be mitigated through a non-discriminatory and verifiable universally acceptable non-proliferation and arms control regime that holds the genuine promise of universal nuclear disarmament.

Notes

1. As quoted in Jonathan Schell, "The Folly of Arms Control," *Foreign Affairs*, vol. 79, no. 5, September/October 2000, p. 38.
2. Ibid., p. 33.
3. SORT is an example of this approach to arms control that forgoes extended negotiations and complicated treaties. The issue has been dealt with in Chapter 8, "Reducing Nuclear Weapons, Increasing Nuclear Dangers."

3

The NPT: An Indian Perspective

Manpreet Sethi

The nuclear Non-Proliferation Treaty (NPT) has never been a favourite with Indians. If anything, it has been India's favourite punching bag, subject to a plethora of criticism. Its discriminatory nature, its inability to check vertical proliferation, its incapacity to prevent the clandestine pursuit of nuclear weapon programmes by the keen non-nuclear weapon states (NNWS), its powerlessness in the face of selective proliferation as indulged in by the nuclear weapon states (NWS), and its failure to live up to the commitment on the eventual elimination of nuclear weapons are the major issues on which the treaty has been decried and rendered unacceptable by India.

Ironically enough, these issues are now beginning to also irk a number of states that subscribe to the treaty. Much of this disenchantment was evident at the sixth NPT Review Conference (RevCon) held in April-May 2000, as also at the three Preparatory Committee (PrepCom) meetings held in the three years preceding the RevCon. The same mood was once again evident at the PrepComs convened in 2002 and 2003 in anticipation of the 2005 RevCon. The NNWS have openly expressed their discontentment at the state of the treaty. Sharp differences of opinion on substantive matters continue, even as procedural issues have proved to be no less thorny.

Delegates at the last RevCon worked hard and long to produce at least some sort of a consensual final declaration that could gloss over the differences for the time being and hold the NPT together, at least till the next quinquennial Review Conference in 2005. Consequently, the last day of the conference in the year 2000 stretched on for forty long hours. It ended only when a final declaration acceptable to each one of the 155 states attending the RevCon (out of a total of 187 that were then members of the NPT) could be achieved. The need for the marathon session provides some indication of the complex nature of issues that animated the discussions. It also manifests the growing rigidity in country positions on matters that member states perceive as impinging upon their national interests. Obviously, then, challenges before the NPT are many at this present juncture, and Ambassador Moher of Canada did not mince his words in bringing this forth:

> There are some very significant challenges in front of this (NPT) regime. I do not think anyone should ever operate on the assumption that a treaty is above question. As long as people see a reasonable return on their investment, they will continue to participate. *If they perceive that the treaty is not living up to their expectations, they may reconsider.*[1]

In a similar vein, Frank Blackaby has written about the prospect of an *en masse* withdrawal from the NPT in case member nations felt that subscription to the NPT was no longer worthwhile.[2]

What do such calls for a reconsideration of the NPT by the treaty members entail for India? A long time critic of the treaty, should it confine itself to mouthing a mocking "I told you so" in the face of a vindication of its objections to the treaty? Or, can it somehow take a more proactive, though subtle, approach to veer the treaty along its own line of thinking? Is the continuation of the NPT, albeit in a slightly amended form, in India's interest or not? These are some of the questions that this chapter endeavours to find answers to. In order to do so, nevertheless, it becomes necessary to first assess the present state of the treaty and examine the nature of the challenges it faces.

The State of the NPT Today

All through the years of its existence, threats to the NPT have always been deemed to be arising from the outside—from nations refusing to subscribe to it, nor accepting its safeguards regime. Efforts, therefore, have traditionally concentrated on ensuring the universality of the treaty. In the years preceding the extension of the treaty in 1995, these attempts were further intensified, so that by the time of the Extension Conference, 175 nations had been brought into the NPT. This was achieved through a judicious use of the carrot and the stick, particularly by the USA. Since then, the numbers have risen further to 188, leaving only three countries out of the NPT fold. These three, namely, India, Pakistan and Israel, are already known to be nuclear weapons capable.

As a result, it should be safe to assume that horizontal nuclear proliferation should not really be a matter of concern any longer. Unfortunately, though, this is not the case. Nuclear proliferation and defence continue to occupy centre-stage, and the US, the most vociferous spokesman for the need to check it, has been most concerned with disclosures such as those of Iraq and North Korea, as also with the suspected nuclear weapon programmes in Iran and Libya. The fact that the US is now vigorously pursuing its plan to deploy a National Missile Defence (NMD) to protect itself from the possibly nuclear tipped missiles of these "states of proliferation concern", provides an inkling to the effect that Washington might have lost faith in the ability of the NPT to stem further proliferation. Why should this be so since these countries that the US describes as being of proliferation concern are NPT members and cannot *legitimately* pursue a nuclear weapons programme?

To a large extent, the responsibility for this should be placed at the doorstep of the NWS themselves since they have done precious little to bring down the role and salience of nuclear weapons. Nuclear deterrence and defence continue to occupy center-stage in the national security strategies of each one of them. In fact, post-1995, each one of the five NWS has individually conducted some sort of a national security review only to arrive at the conclusion that notwithstanding the end of the Cold War, enough threats remain or are likely to emerge that shall warrant the need to retain their nuclear arsenals into the indefinite future. Obviously, then, the message that filters down to the rest of the international community is that nuclear weapons are useful tools for security and status, and if national interests so deem it necessary, it would be worthwhile to acquire them, NPT membership notwithstanding.

Many of those nations that have so felt the need have found it easy to subvert the relevant provisions of the NPT and its safeguards regime or other export controls. In fact, in many cases they have even been helped in the process by the policy of selective proliferation of nuclear and missile technologies that the NWS have been guilty of. While in some cases, the NWS have, in violation of Articles I and II of the NPT, passed on nuclear weapons or technology to other NNWS, in others they are equally guilty of condoning others' activities of this kind. China, for instance, has never let itself be constrained by its NPT membership and has continued to indulge in widespread transfer of WMD (weapons of mass destruction) technologies to Pakistan, North Korea or Iran. The US and its other nuclear weapon allies, however, have never thought it prudent to rap the Chinese knuckles on this count.

Actions such as these have undermined the credibility of the treaty in the eyes of those who have striven to genuinely uphold their part of the bargain by refraining from acquiring nuclear weapons. But can this restraint be expected to last forever, since the role and importance of nuclear weapons and nuclear deterrence are on the ascendant? The challenge, then, for the NPT presently arises from within, from the actions and overtures of its own member states—both nuclear and non-nuclear. Despite its near universality, the major threats to it arise from the widening fissures within the ranks of NPT subscribers.

The risk of disintegration of the NPT because of implosion is further compounded by a number of issues on which differences in perception and priorities of countries have only been increasing. Some of these contentious issues, both procedural and substantive, are explained in the following section.

Issues Likely to Derail the NPT

Procedural Problems

The sixth RevCon, in a way, signified a sort of test case since this was the first review being held after the treaty's extension in 1995. At that time, the indefinite extension of the NPT without a vote had been secured by granting three concessions. They are like sops that the NWS held out to the NNWS in order to pacify them and relieve them of the troubling thought that by granting an indefinite extension, they were surrendering their leverage over treaty implementation with regard to Article VI that specifically deals with the pursuit of efforts "in good faith" for the eventual elimination of nuclear weapons. The three decisions were held out as a means to secure the NPT's "permanence with accountability". However, as the RevCon demonstrated, accountability is still amiss and the permanence of the treaty could also be in jeopardy!

The first of the three decisions sought to strengthen the review process through an enhanced and more substantive review of treaty implementation. Besides assessing progress in the past five years, the review process was also given a forward-looking orientation by tasking it with the identification of steps for the future. And, to carry out this exercise, three PrepCom meetings were envisaged in the last three years before the run up to the

RevCon. The task of the PrepComs was also expanded beyond the traditional mandate of considering only procedural matters to include discussions and recommendations on substantive issues as well. As the Resolution specified, "...purpose of the Preparatory Committees would be to consider principles, objectives and ways in order to promote the fullest implementation of the treaty, as well as its universality, and to make recommendations thereon to the Review Conference...These meetings should also make the procedural preparations for the next Review Conference..."[3]

Decision No. 2 defined a list of Principles and Objectives for Non-Proliferation and Disarmament. These were to be the benchmarks against which it was to be assessed as to how much, and in which direction the international community, especially the NWS, had moved on these two crucial issues. It specified guidelines and indicative targets designed to promote greater accountability.

The third resolution related to the Middle East and, as is pretty much self-evident, was included to garner the support of Egypt and the Arab states for the NPT's indefinite extension. In 1995, none of these countries was in favour of an indefinite extension, not even Kuwait and Saudi Arabia that were perceived as client states of the US after the 1990 Gulf War. The resolution on the Middle East was then worked out to rope them in. It expressed concern on the continued presence of unsafeguarded nuclear facilities in the Middle East and called for "the establishment, by regional parties, of a Middle East zone free of nuclear and all other weapons of mass destruction and their delivery systems."[4]

The three decisions constituted a package that would appeal to the maximum number of NPT member states in order to secure its indefinite extension. This was the priority of the NWS in 1995 so that their right to retain, in perpetuity, their monopoly on nuclear weapons would stay untouched. Working on this with single-minded devotion, they might not have thought much of granting these concessions, nor anticipated the kind of trouble they were getting themselves into by legislating a Strengthened Review Process (SRP) that would have the power to actually scrutinise their actions against a set of principles on non-proliferation and disarmament.

An examination of the three PrepComs and the sixth RevCon reveals the widespread implications of the three decisions. As mandated by the

SRP, the PrepComs held in 1997, 1998 and 1999 examined and debated substantive issues besides considering procedural matters. In the process, and contrary to the expectations and wishes of the NWS, each one of these meetings turned out to be akin to a mini Review Conference. The NNWS perceived the PrepComs as a forum through which to make the NPT review a more credible and meaningful process for ensuring treaty accountability. Consequently, the NWS found their nuclear policies and decisions subjected to minute scrutiny and comment.

In a turn of events not anticipated by the NWS, each one of the meetings turned out to be an acrimonious affair, generating more heat than light. Initially, it had been thought that the first PrepCom would start a rolling text covering the subject matter of the treaty. This was then to be worked upon by each successive PrepCom so that by the end of the third PrepCom, a set of agreed recommendations would be ready to be placed before the RevCon. However, no such thing happened as the very issue of having a rolling text became embroiled over the kind of recommendations that could be made and the status that they should be accorded. While some considered that the recommendations put forth by the PrepComs would be binding and have to be included in any RevCon outcome, others argued against it, upholding the independence of the RevCon.

At the very first PrepCom itself in 1997, differences of opinions and interpretations arose on such basic issues as the role and function of the PrepComs—on whether they should focus on information exchange, or on implementing the Principles and Objectives from the previous RevCon and provide an annual report based on that, or develop a text for the RevCon, or perform all the three functions. Interestingly, the attendance at this meeting was impressive with representatives from 149 states and 135 NGOs (non-governmental organisations). It was also illustrative of the hope and optimism at the potential of the SRP. However, the meeting could not live up to the expectations. Rather, it exposed differences between groups of nations on the very meaning and scope of the SRP, and owing to a lack of consensus, the first PrepCom failed to forward any major recommendations to the second one.

That disillusionment with the SRP was beginning to set in was evident in 1998 when only 97 nations and 76 NGOs found it worthwhile to attend the meeting, in contrast to the impressive numbers of the previous year.

In addition to the differences that still persisted, the second PrepCom encountered another procedural hurdle over the rule covering the work of committees. While South Africa, backed by the countries of the Non-Aligned Movement (NAM) and some other NNWS wanted the mention of working groups to be supplemented by an explicit reference to subsidiary bodies, Russia objected to it. Attempts to include both terms also foundered. Finally, with time running out, the second PrepCom passed on all its problems to the next one.

The third and final PrepCom managed to register a few procedural and organisational advances. For instance, it was agreed that subsidiary bodies on nuclear disarmament and on regional concerns, with reference to the Middle East, would be established to look at the issues in a focussed manner. This signified a major turnaround of the US and Russian positions since they had vehemently opposed the concept of subsidiary bodies at the second PrepCom. However, the two countries sought to downplay their change in positions by describing it as a mere change of name. As the Working Paper presented by the US at the third PrepCom stated, "The change of name does not imply any change in the status of these subsidiary bodies."[5]

In another significant compromise, the US agreed to allow the UN Secretariat to prepare background documentation on the Middle East. At the second PrepCom, Egypt and other Arab states had maintained the provision of background documentation on the issue and insisted that the NPT must exhort Israel, by name, to join the treaty. The US had then resisted these attempts and argued that the Middle East resolution agreed to in 1995 was a one-off stand-alone agreement that had no validity beyond that date.[6] In fact, the breaking point for the 1998 PrepCom came over this issue. While the 14 Arab states, backed by the NAM, requested for background documentation, the US refused to comply, holding that background documentation should be limited to addressing treaty Articles only. It opined that though the resolution had been agreed to at the 1995 Extension Conference, it was inappropriate to have documentation on an issue not referred to in the treaty itself.[7] Therefore, the US had refused to accept the reference to the Middle East in the Canadian paper, even though the language did not go beyond the 1995 Middle East resolution. Meanwhile, the NAM asserted that if this was not passed, it would not agree to any substantive item, and Egypt and Iran made it known that the acceptance of the Middle East language was the unmovable bottom line.

By the third PrepCom, however, the US had agreed to allow background documentation on this issue. It conceded that the resolution formed an integral part of the 1995 package and that it would remain valid until those goals were achieved.

The changes in the position of the US, as also of some other NWS, may have been prompted by a realisation that unless the PrepComs met with some success, disenchantment with the treaty would only grow and this would not augur well for the future of the treaty. Therefore, some flexibility, at least on procedural matters, was found prudent so that delegates leaving New York City after the final PrepCom could return the next year in a more positive frame of mind. Yet it cannot be denied that the inability of the PrepComs to address procedural issues in any meaningful way, thereby belying the original intention of the SRP, cast a shadow over the RevCon. The feeling of scepticism and frustration at the inability to transact any meaningful business added to the pressure to register some sort of a success at the RevCon, especially so that the future of the NPT may not be jeopardised. As the president of the conference, Ambassador Baali of Algeria said, "The outcome of this Conference will have a major impact on deciding the future course of the NPT and the nuclear non-proliferation regime for generations to come."[8] Perhaps, it is also for this reason and sensing just such a mood that the state parties strove to at least bring out a consensual final document as a token of the success of the NPT.[9]

The P-5 Statement

Apart from the adoption of the final document, the other major achievement of the sixth RevCon was a joint P-5 statement that was put forth in the second week of the RevCon. In a clear bid to restore confidence in the NPT, the NWS issued a statement committing themselves unequivocally to the ultimate goals of a complete elimination of nuclear weapons and to the early conclusion of a treaty on general and complete disarmament. This was the first such joint statement at an NPT RevCon.[10] While collectively reiterating the significance of the Comprehensive Test Ban Treaty (CTBT) and the early conclusion of the Fissile Material Cut-off Treaty (FMCT), the five NWS promised to provide individual accounts of their systematic and progressive efforts at reducing nuclear weapons globally. In another significant paragraph of the statement, they declared that none of their nuclear weapons was targetted at any state. While retargetting can be done quickly, this has been hailed as a major confidence building measure (CBM).

Apart from the declaration on detargetting, no major progress from positions that already existed may be inferred from the P-5 statement. The unequivocal commitment hardly implies anything more than what has been enshrined in Article VI of the treaty for 32 years and the commitment remains as ambiguous and open-ended. The British defence minister even said as much in a BBC interview shortly after the issuing of the joint statement. He said the pledge was an "agreement in principle to which there was no time-table attached."[11] So it has remained ever since the NPT first came into being. Rather, the P-5 statement once again put nuclear disarmament back into the context of total worldwide disarmament—a linkage that the advisory opinion rendered by the International Court of Justice (ICJ) in 1996 had denied.[12]

Perhaps, the only major significance of the P-5 statement lay in the fact that the NWS felt troubled enough by the drift within the NPT to put aside their differences to issue a joint statement. They sensed the mood of the NNWS and, hence, perceived a need to bring out this statement. By issuing it at the time that they did, the NWS sought to ensure that the treaty's future would not be jeopardised, at least not for another five years and till the next RevCon. At the same time, it sought to unify the five NWS themselves, since fissures were most evident over the possibility of a revision of the Anti-Ballistic Missile (ABM) Treaty and the Chinese intransigence over the issue of the FMCT. It was as if the members of the nuclear club found greater merit in sticking together to keep the treaty alive and thereby retain their privileges. They realised, perhaps, that united they could withstand the mounting pressure of the NNWS to make them give up their nuclear weapons. In any case, whatever may have been the motive of the P-5 statement, to some extent, it may be credited for having had a positive impact on attempts to reach a final document. This becomes doubly important when one realises that the RevCon was likely to be stalemated over a number of issues and these are examined in the following paragraphs.

Nuclear Disarmament

The experiences of the three Prepcoms and the RevCon that followed illustrate the extent of importance that countries, especially the NNWS, have come to attach to the issue of nuclear disarmament. At the same time, the deliberations also reveal the level of frustration that countries now feel at the lack of intention and willingness on the part of the NWS to live up to their part of the bargain in the NPT. Under the provisions of the treaty, the NNWS committed to remaining non-nuclear and the NWS undertook to work towards the eventual elimination of their nuclear arsenals. While a majority of the NNWS have kept their word, the NWS have hedged on their share of responsibility and post-1995, when the treaty was extended indefinitely, it has been felt that the NWS are unlikely to ever give up their monopoly over nuclear weapons.

The pursuit of the goal of universal nuclear disarmament, hence, appears to be bogged down by the divergent ways in which the issue is perceived by different states. On the one hand, the Western NWS tend to take recourse in recounting the progress that has already been made in this direction by playing up START I and II and the CTBT, etc. The US, for instance, has never tired of pointing out that it has reduced 90 per cent of its non-strategic nuclear stockpile and 47 per cent of its strategic nuclear stockpile from what it was at the peak of the Cold War.[13] France, too, on behalf of the other NWS, has highlighted the important and tangible progress achieved so far and vaguely pledged to continue "systematic and progressive efforts."

For further progress, the NWS insist on regional stability as a precondition for nuclear disarmament. The US has made it clear that *"nuclear disarmament will not take place on demand, but could only come about as a consequence of conventional disarmament and regional stability."*[14] Therefore, it has been exhorting the NNWS to remain realistic about the pace of progress in the near term. France, too, has stated, "It is the responsibility of all states to contribute to the relaxation of international tension and to the strengthening of international peace and security." In fact, it is very interesting to note that each of the five NWS has conditioned further progress on this issue on matters that are considered crucial to their security. For instance, the US warned, "If countries demand unrealistic and premature measures, they will harm the NPT and set back everyone's cause". Russia too voiced its opposition to granting too much importance to the attainment of nuclear disarmament. At the second PrepCom in 1998, the Russian delegate stated, "We believe it is untimely, and consequently, counterproductive to start talks at the Conference on Disarmament (CD) on a programme for nuclear disarmament within specific time limits."[15] Since then, Russia's position has further hardened on the subject, since at the RevCon, Moscow linked further progress on nuclear disarmament to the continuance of the ABM Treaty. As is well known, Russia was opposed to the deployment of NMD by the US through any revision of the ABM Treaty. Apprehensive of the likely impact it would have on its own nuclear deterrent, it had conditioned all further actions on nuclear disarmament on the continuance of the ABM Treaty.

In fact, on this issue, Moscow has found an ally in China. Traditionally a supporter of nuclear disarmament, China has taken a harsh view of what it deems are US designs to eventually militarise outer space. Therefore, it has predicated its participation in arms control measures on two conditions—that they do not compromise global strategic balance and stability; and, that they do not undermine China's national security interests. Since China will be the sole determinant of whether these conditions are satisfied or not, its support to nuclear disarmament appears to be little more than a case of being politically correct. France, on the other hand, has agreed to discuss issues relating to nuclear disarmament at the CD provided a mandate can be adopted by consensus. A tall order, given that differences run deep with the other camp.

The other camp primarily comprises the NNWS that deny that nuclear disarmament could be achieved only when general and complete disarmament has been accomplished.[16] For instance, Canada, supported among others by Japan, China, South Africa and New Zealand, has questioned this contention. In a strongly worded statement, the Canadian representative said, "We do not accept any explicit or implicit linkage, or interpretation of Article VI, that nuclear disarmament will be achieved only when general and complete disarmament has been achieved, or when every last bow and arrow or Swiss Army knife is gone."[17]

Other NNWS too have been pushing for concrete movement on nuclear disarmament, though not entirely in a unified manner. For instance, the NAM has sought negotiations for the complete elimination of nuclear weapons within a specified time-bound framework. Ireland brought into focus that the PrepCom should not ignore "external opinion" on nuclear disarmament, such as that of the ICJ, the Canberra Commission, or the Programme of Action tabled by a group of 28 nations calling for elimination of nuclear weapons by 2020. Mexico too supported this viewpoint. It categorically insisted on adding nuclear disarmament to the list of recommendations to be presented by the first PrepCom to its successor on the ground that the 1996 Advisory Opinion of the ICJ held the threat or use of nuclear weapons to be in contravention of international and humanitarian law.[18]

Meanwhile, the New Agenda Coalition (NAC)[19] encapsulated its position and suggestions in a working paper presented at the RevCon. The NAC list of actions included a call for reexamining nuclear doctrines, removing nuclear weapons from delivery vehicles, eliminating tactical nuclear weapons, negotiating a global legally binding negative security assurance treaty, and holding a dedicated conference on nuclear disarmament. Above all, it held that actions in the direction of nuclear disarmament should be transparent, irreversible and accountable. The NAC proposals have garnered increasing support from among the NNWS, but, naturally, the NWS are opposed to a majority of the NAC suggestions, especially those seeking some sort of a time-bound programme of action.

The UK, on behalf of all the Western nuclear powers, made it clear that "to re-focus now on rigid time-frames or single instruments would be to jettison a tried and tested method for moving towards our goal in order to repeat the errors of the past."[20] Russia too voiced its opposition to granting too much importance to the attainment of nuclear disarmament. The PrepCom was informed by the Russian delegate, "We believe it is untimely and, consequently, counterproductive to start talks at the Conference on Disarmament on a programme for nuclear disarmament within specific time limits. We are convinced that it is not the right time to be seriously engaged in something that could even hardly acquire practical grounds during the first decade of the next century."[21]

Such divergent opinions on nuclear disarmament spilled out into the open at the RevCon. The main committee[22] looking at this issue, broke it up into two parts—one dealing with unfinished business such as the CTBT, FMCT, START, etc., and the other focussing on the identification of steps for the future. Expectedly, there wasn't any consensus on the latter. While the NAM and NAC sought acceleration in negotiations between 2000-05, Britain, the US and Russia were loath to confining themselves to a time-frame. Russia further expressed its reservations on the clauses calling for irreversibility and those addressing the issue of tactical nuclear weapons unless they were put in the context of strategic stability. In fact, the term strategic stability became a sort of buzzword that the NWS exploited whenever necessary. It may be taken to mean stability made possible by the mutual retention of comparable levels of nuclear weapons. And each time they were faced with a proposal that appeared to be threatening their hold over their nuclear arsenals, the NWS resorted to the tactic of "agreeing with the proposal in principle, subject to its being able to maintain strategic stability."

The final document put forth by the RevCon, of course, did not really break any new ground on the crucial issue of nuclear disarmament. It confined itself to expressing deep concern over the continued threat to humanity from the possession of nuclear weapons by some states and over the reaffirmation of nuclear weapons doctrines, including those that retain first use. In the paragraphs looking ahead, it urged the NWS to make further efforts at reducing nuclear arsenals unilaterally, and accepting the principles of irreversibility and transparency as a voluntary measure since they could go a long way in confidence building. Expectedly, and rather timely, it also sought an early entry into force of the CTBT, negotiations on an FMCT at the CD, and the establishment of a subsidiary body in the CD to deal with disarmament.

If the NWS continue to exhibit the same callousness and casual approach toward their commitment to Article VI of the NPT, then disenchantment with the NPT can only be expected to grow. In this context, it would be instructive to recall the perceptive words of Ambassador Jayantha Dhanapala (president of the 1995 NPTREC):

> If there is naked cynicism on the part of the nuclear-weapon States and a total disregard of nuclear disarmament commitments...then we might see not just one or two countries for individual reasons wanting to opt out...but a major threat of an exodus from the treaty using Article X.1 as an exit clause...is a very grave danger. We must never ever let the Treaty be in jeopardy, and for that there has to be progress in nuclear disarmament.

Nuclear Sharing

Another related issue that has been vociferously raised by the NAM at the NPT meetings is that of the nuclear sharing arrangements under NATO (North Atlantic Treaty Organisation). All through the three PrepComs, the NAM presented working papers calling upon the NWS to refrain from nuclear sharing for military purposes "under any kind of security arrangements among themselves, with non-nuclear weapon states, and with states not party to the Treaty."[23] It may be mentioned that even though the NPT prohibits the NNWS from acquiring nuclear weapons and the NWS from transferring them, six non-nuclear NATO member states (Belgium, Germany, Greece, Italy, Netherlands and Turkey) house American nuclear weapons on their territory.

Notwithstanding these demands, however, NATO issued its new Strategic Concept in April 1999, only a month before the final PrepCom was scheduled to meet. It reaffirmed its commitment to nuclear sharing and the first use of nuclear weapons. The new document made it explicit that "nuclear weapons make a *unique contribution* in rendering the risks of aggression against the Alliance incalculable and unacceptable. Thus, *they remain essential to preserve peace.*"[24] The US maintains that NATO remains the anchor of American engagement in Europe and within that "the presence of US conventional and nuclear forces in Europe remains vital to the security of Europe, which is inseparably linked to that of North America."[25]

Given the articulation of such positions at the NATO Summit, it was hardly surprising that the third PrepCom witnessed angry outbursts on the issue of nuclear sharing. Algeria criticised the Concept for reaffirming the essential importance of nuclear weapons, thereby contradicting by word and deed the hopes of eventual nuclear disarmament cherished by so many countries. Mongolia too cautioned that the NATO action could provoke other NWS to adopt similar policies and question the very utility of the NPT.[26] Egypt recommended that the NPT should be interpreted in a way so as to outlaw current NATO practices and possible future EU (European Union) nuclear weapons cooperation.[27] Drawing upon these sentiments, at the sixth RevCon, the NAM sought to put the removal of US nuclear weapons from European territories as the next step towards disarmament. However, its attempt came to naught since the final document does not include any reference to this.

Nevertheless, the frustration and dissatisfaction that has made itself evident on this issue indicates that it is likely to remain a live one in the years to come.

Iraq's Non-Compliance

Yet another subject that kept the RevCon in a state of animated suspension right until the end was that of Iraq's non-compliance with safeguards. In his report to the conference, the International Atomic Energy Agency (IAEA) director general stated the inability of the IAEA to provide any assurance that Iraq is in compliance with its obligations. But the Iraqi representative dismissed this and said his country was in full compliance with all its obligations under the NPT and the IAEA safeguards regime. Rather, in a bid to divert attention from his own country, he protested that no action had been taken to force Israel to join the NPT or to place its facilities under safeguards.

Therefore, at the stage of drafting the final document, a problem arose over the manner in which to make a mention of this issue. A stand-off on the subject between the US and Iraq at one time seemed likely, thereby threatening to scuttle the possibility of a consensual final document. A compromise was subsequently reached when Iraq conceded that its name may be mentioned for non-compliance, provided the final document also noted that in January 2000, after Iraq's conclusion of a safeguards agreement with the IAEA, inspections were carried out and were able to verify the presence of nuclear material subject to safeguards. The conference reaffirmed the importance of Iraq's full and continuous cooperation with the IAEA.

While the matter may have been resolved for the moment at the sixth RevCon, the point that must be noted for the future is the possibility of other countries seeing merit in Iraq's behaviour and being tempted enough to follow the example. Unfortunately, the manner in which the issue was resolved casts a shadow over the efficacy of the NPT. It points to the fact that accession to the NPT and the document of IAEA safeguards can prove to be no guarantee against a country's attempts to attain nuclear weapons. They cannot prevent a determined proliferant who sees some value, either in terms of enhanced security or status, in acquiring them. And, this value would diminish only when nuclear weapons are eliminated from the arsenals of all countries.

Therefore, the Iraqi case holds an important lesson that India must seek to highlight. It has to make the point that denuclearisation cannot be dictated in a selective and isolated manner as it has been sought in the case of Iraq. In doing so, rather, not only has no success been achieved, but it has also resulted in projecting nuclear weapons as a powerful bargaining chip. Therefore, if at all denuclearisation has to be imposed, it could work only when it is done through a universally applicable and non-discriminatory mechanism.

Israel's Non-Adherence

Moving on from the problem of non-compliance to non-adherence, the mention of Israel crept into the RevCon discussions several times. As in the past, Egypt took the lead in pushing the NPT to address the problem posed by Israel's unsafeguarded nuclear facilities and nuclear weapons capabilities. It also presented a working paper with a proposal that envisaged:

(a) The setting up of a follow up committee comprising the chair of each session of the PrepComs plus the three depositary states[28] to initiate contacts with Israel and report back to successive RevCons.

(b) Designating a special envoy from the NPT states to pursue discussions with Israel and report back.

The proposal managed to garner the support of the NAM and Arab countries. But the US was opposed to singling out Israel in such a manner. Rather, it drew attention to the need for progress in the broader peace process, which would naturally improve prospects for a nuclear weapon-free zone (NWFZ) in the region. The UK, meanwhile, objected to the depositary states having any special responsibilities on this count. It quoted the 1995 Resolution as calling upon *all states parties to the NPT* to work for an NWFZ in the region. Finally, the RevCon final statement reaffirmed the importance of Israel's accession in realising the goal of universality. It also requested all states parties, particularly the NWS and states of the Middle East and other interested states to report through the UN Secretariat to the president of the next RevCon and to the PrepComs on the steps taken to promote an NWFZ in the Middle East.

The accession of Israel to the NPT is part of the same problem as that of India and Pakistan. While the latter two countries are declared nuclear weapon states, Israel has refrained from making any official announcement to this effect though it has never left anyone in doubt of its nuclear weapons capability. The NPT closed the doors to new NWS on January 1, 1967, and hence, any future subscription of these three countries poses a unique dilemma. They cannot be accepted as NWS and they cannot be expected to roll back or eliminate their programmes to make themselves available as NNWS. The significance of whatever solution could be arrived at is heightened by the fact that it shall be some sort of a precedent and spell repercussions for the entire NPT community. However, this is an issue that it may be called upon to address in the forthcoming reviews, especially in the context of Israel.

Export Controls

Another subject that was considerably debated at the RevCon was that of export controls. Several NAM countries expressed concern at the way export controls hindered their right to participate in the "fullest possible exchange of equipment, materials and scientific and technological information" provided for in Article IV of the NPT. The NAM Working Paper argued against any denial on the basis of allegations of non-compliance not verified by the IAEA. Understandably, Iran expressed its dismay over the systematic denial of transfer of technology and criticised restrictive export control policies.

The nuclear suppliers, however, pointed towards the large degree of transparency on export controls that had been achieved since 1995. Currently chairing the Nuclear Suppliers Group (NSG), Italy gave examples of the measures the NSG had taken to increase transparency and its claim was supported by many, including Australia, the USA and the EU countries.

In the final RevCon statement, the issue of export controls was completely watered down or lost. Significantly, China refrained from committing itself to making full scope safeguards a condition for nuclear supplies. Does this have any implications for its nuclear relationship with Pakistan? Also, interestingly, Russia and France too were ambiguous on this subject, and does that signify anything for the fact that both are desirous of finalising nuclear deals with India?

Conclusion

The experience at the recent meetings of the NPT indicates that a sense of cynicism is beginning to pervade the treaty members. While the delegates at the sixth RevCon deliberated long and hard and emerged from it with relief writ large on their faces, it is debatable whether everyone felt an equal sense of satisfaction at what the review achieved. In an attempt to save the treaty from unravelling, each one of them walked a tightrope, faced with the twin challenges of making the conference a success by producing a consensual final document, and at the same time making sure that their national or bloc positions were not fundamentally compromised.

The task could not have been easy since every nation raised issues and concerns and made statements reflecting its own security interests. For instance, while no one deemed it politically correct to oppose negotiations for an FMCT, country positions, nevertheless, exposed the seriousness, or lack of it, that they attached to it. In this context, the stance taken by China is demonstrative. It made negotiations on a fissban conditional on the CD working out a programme of work that would be balanced and comprehensive. And this, according to the Chinese delegate was possible only if the CD held concurrent negotiations on the issues of outer space and nuclear disarmament. Apprehensive of the US deployment of nuclear weapons in outer space under the guise of the NMD, Beijing considers this even more urgent than the conclusion of an FMCT. In fact, it did not agree to even call for a collective moratorium on the production and stockpiling of fissile material, even though it is widely believed to have stopped production of fissile material some years ago. Rather, in an intelligent move, China linked the call for a moratorium on fissile material production with one seeking a moratorium on deployment of weapons in outer space.

Similarly, if one looks at the issues that Russia raised at the RevCon, they originate from its own immediate threat perceptions. It considers local conflicts, international terrorism and militant separatism as the factors fuelling nuclear proliferation. In an oblique reference to the US NMD, it also decried the attempts of states to undermine the system of strategic stability.

While the NWS harped on the dangers from horizontal proliferation and, hence, the need for universality and stringent supporting safeguards, the NNWS remained more concerned with gaining access to nuclear technology and disarmament. As is evident from this, the RevCon worked in an atmosphere in which parties pushed for implementation of the elements they found desirable while resisting those not in their interests. And in doing so, they resorted to the use of a beautifully clever play of words. For instance, China accepted nuclear disarmament, but also said that the measures should follow the principle of maintaining global strategic stability and undiminished security of every state. France, meanwhile, blocked a paragraph enshrining the unequivocal undertaking to accomplish the total elimination of nuclear arsenals. Instead, all it was ready to accept was an unequivocal undertaking to accomplish their ultimate elimination. Meanwhile, the phrase strategic stability was evoked in every context, from reducing nuclear weapons unilaterally, to dealing with tactical nuclear weapons, to reducing the role of nuclear weapons in security policies.

Stalemates on several issues were encountered. But meetings ending in deadlocks too often surely lower the credibility of the treaty and dilute compliance with it. Should some states fail to faithfully live up to their commitments, it would be only natural to assume that others too would be tempted to take their obligations lightly. This is the challenge that the NPT was to guard against at the present moment. In fact, despite its near universality, proliferation could pose an even greater threat than earlier precisely because the temptation to proliferate has been heightened by the events of the past couple of years. A feeling of frustration with the lack of progress on NPT commitments could push countries into reviewing the usefulness of their continued subscription to the treaty, especially since there seems to be no hope of its ever achieving the objective of complete nuclear disarmament. It is in this context that the NPT review processes would have to not only perform their traditional role of reviewing the past year and looking forward imaginatively, but also do so in a manner to restore the faith of the members in the efficacy of the review process itself. In the absence of steps being taken to restore their faith, countries might be compelled to look elsewhere for solutions.

What could be the nature of this reconsideration and what options could it throw up? Having mandated its indefinite extension, do the NNWS enjoy any leverage to get the NWS to consider their demands? There are some options that could be considered in this regard. For instance, the NNWS do have a right to call for an amendment conference. Under Article VIII of the NPT, any state party to the treaty can request for such a conference and then, if the demand is supported by one-third or more of the states, the depositary states would be obliged to convene an amendment conference.[29] At this meeting, then, they could propose to consider a concrete programme for the elimination of nuclear weapons. Secondly, the NNWS have the option of withdrawing from the treaty. In fact, this has begun to be thrown up as a possible means of effectively influencing the NWS into becoming serious over the realisation of an NWFW (Nuclear Weapon Free World).[30]

While there is nothing to suggest that the NNWS have yet reached a point where they are going to exercise any one of these options, the general ambience of discontent at the review meetings does not discount such a possibility in the future. It is this sense of dissatisfaction that India must seek to exploit in order to have the treaty amended along lines that it feels would be more favourable to its own interests, especially with regard to the pursuit of universal nuclear disarmament that has been India's long-standing goal. Within the NPT, increasingly practical proposals are now beginning to be put forth by Canada, Mexico, South Africa and the NAC. Many of the steps outlined in the NAM and NAC Working Papers on the subject mirror those that India has consistently put forth in its own resolutions to this effect before the UN General Assembly (UNGA). Hence, India could let it be known that it explicitly supports movement in this direction. It must be understood that notwithstanding its shortcomings, there is widespread consensus that the NPT and its continuance is in the interest of every state. And one could venture to add that it would even be in the interest of India. Through its verification mechanism and international safeguards, the NPT does put brakes on rampant nuclear proliferation, and to have less nuclear weapon capable states is in India's interests. As a next step now, in case the NWS could be made to agree to a clear and time-bound plan for the elimination of nuclear weapons, irrespective of the schedule in terms of the number of years, the Indian objection regarding the discriminatory nature of the treaty would eventually vanish. A reconsideration of the treaty could then be possible. Especially so, since the problem of the status under which countries like India and Pakistan should be granted membership of the treaty would also be obviated.

There were several who, at the time of the extention of the NPT in 1995, had described it as the high point of global efforts to prevent nuclear weapons proliferation and move towards global disarmament. Indeed, while the 1995 Conference may have technically and legally secured the treaty's indefinite extension, the 2000 RevCon actually tested whether such an extension can guarantee permanence for the treaty in the wake of a host of adversely impacting developments. The next RevCon in 2005 will prove to be even more crucial since it will serve as the true barometer of the state of the NPT.

In fact, as the first conference to be held post-September 11, a major watershed in international threat perceptions, the 2005 RevCon shall be put to an even more severe test. Indications of what could happen in three years from now were evident at the 2002 PrepCom. While most countries expressed the view that in the current international climate wherein security and stability are being severely challenged, both globally and regionally, by the proliferation of WMD and their means of delivery, preserving and strengthening the NPT is vital to peace and security, many others opined that the terrorist attacks on September 11, 2001, had given an even greater sense of urgency to the efforts of all states in the field of disarmament and non-proliferation. The view was held that further strengthening and reinforcing the non-proliferation regime was imperative to prevent the use of nuclear materials and technologies for criminal/ terrorist purposes.

At the same time, there were also signs of deep concern, not expressed though in open interventions, that the Bush Administration's behaviour towards multilateral treaties and regimes as in the case of the CTBT, ABM Treaty, Kyoto Protocol, etc. was not encouraging for the NPT. Moreover, the implications of the classified but leaked Nuclear Posture Review (NPR) cannot but have to be factored into the implications for further proliferation.[31]

Therefore, the predominant sentiment at the last Prepcom had a distinct undertone of resignation. While the NPT member states may continue to go through their routine diplomatic moves at the future PrepComs too, exhibiting a lethargy to rock the non-proliferation boat, the lack of enthusiasm and gradually eroding faith in the treaty can surely not augur well for non-proliferation in the years to come. The NPT needs some serious rethinking and the forthcoming RevCon must take on the challenge.

Notes

1. Ambassador Mark Moher, the Canadian permanent representative at the CD as quoted by Stephen Young, "1999 NPT PrepCom: Keys to Success," *Occasional Papers on International Security Issues,* BASIC, no. 30, April 1999. Emphasis added.
2. Frank Blackaby, "Time for A Peasant's Revolt," *Bulletin of Atomic Scientists,* September-October 1997.
3. NPT/CONF. 1995/32 (Part 1), Decision 1, Para 4.
4. Resolution on the Middle East reported in the Final Document of the Conference as NPT/CONF.1995/32/RES.1, May 11, 1995.

5. Working Paper submitted by the USA, NPT/CONF.2000/PCIII/6, May 10, 1999.

6. Rebecca Johnson, "The 1999 PrepCom: Substantive Issues", *Disarmament Diplomacy,* no. 36, April 1999, http://www.acronym.org.uk/36reb.htm

7. The USA has maintained that the resolution on the Middle East was a "one-off stand alone resolution" and not part of the package of decisions to extend the treaty.

8. Statement of the president of the 2000 RevCon, H.E. Ambassador Abdallah Baali, New York, April 24, 2000, http://www.basicint.org

9. Interestingly, the consensual adoption of a final declaration has been possible in only three out of the six RevCons of the NPT held so far: at the first RevCon in 1975, the third one in 1985, and the most recent one in 2000.

10. For the text of the statement, see *Strategic Digest,* June 2000. At the first and second PrepComs though, joint statements by the P-5 had been made. However, a fresh statement could not be made in 1999 because of the deterioration in the US' relations with Russia and China following the NATO action in Kosovo.

11. Thomas Abraham, "Britain Has no Plans to Reduce Stockpile Now," *The Hindu,* May 23, 2000.

12. For a detailed analysis of the ICJ opinion on Legality of the Threat of Nuclear Weapons, see the special issue of *International Studies,* 1998, Jawaharlal Nehru University, New Delhi.

13. Ibid.

14. Douglas Roche, "An Analysis of the First PrepCom Meeting for the 2000 Review of the Non-Proliferation Treaty," http://www.watser1.uwaterloo.ca/~plough. Emphasis added.

15. Douglas Roche, "An Analysis of the Second PrepCom Meeting for the 2000 Review of the Non-Proliferation Treaty," http://www.watser1.uwaterloo.ca/~plough

16. Of course, it may not be understood that the NNWS project a monolithic viewpoint, agreeing on the means and the steps to achieve nuclear disarmament. Differences in how to approach the issue abound. For details on these, see Manpreet Sethi, "The NPT Review Process: Turning on the Heat," *Asian Strategic Review,* 1998-99 (New Delhi: IDSA, 1999), pp. 168-185.

17. Roche, n. 14.

18. It needs to be mentioned that in its testimony before the ICJ, Mexico had warned that it was not ready to accept a monopoly in the possession of nuclear weapons and that if the nuclear disarmament obligations of the NPT were not met, it would "need to revise our continuation as party to the Treaty."

19. The New Agenda Coalition is a grouping of seven countries—Brazil, Egypt,

Ireland, Mexico, New Zealand, South Africa and Sweden. It was formed in 1998 after the failure of the second PrepCom to achieve any progress on substantive matters.

20. Roche, n. 15.

21. Ibid.

22. The RevCon normally consists of a Plenary and three Main Committees for negotiating the key issues of disarmament, safeguards and peaceful uses of nuclear energy. The main committees submit their results to a drafting committee, which then seeks to harmonise the results of all the three into a final document.

23. NAM Working Paper as presented to the RevCon and reproduced in *Strategic Digest,* June 2000.

24. "The Alliance's Strategic Concept," NATO Press Release, NAC-S (99) 65, April 24, 1999 as reproduced in *Strategic Digest,* June 1999. Emphasis added.

25. Ibid.

26. "NATO Nuclear Policies Slammed at the NPT PrepCom," *PENN Press Release,* May 11, 1999 http://www.basicint.org

27. "Egypt Proposes Ending NATO Nuclear Sharing," PENN Press Release, May 12, 1999. http://www.basicint.org

28. The three depositary states are the US, UK and Russia (formerly the Soviet Union).

29. This aspect has been brought out by Zia Mian and M.V. Ramana, "Diplomatic Judo: Using the NPT to Make the NWS Negotiate the Abolition of Nuclear Weapons", *Non-Proliferation Review* (Monterrey Institute, 1999).

30. Blackaby, n. 2.

31. This subject has already been dealt with in some detail in the introduction.

4

India's Stake in the CTBT

Manpreet Sethi

With President George Bush Jr. at the helm of affairs in Washington and given his own and his party's aversion to the treaty, it is highly unlikely that the CTBT in its present form shall enter into force anywhere in the near future. Nevertheless, it might be too early to write an obituary for the treaty. It is only comatose yet, and efforts to revive it at some point in the future cannot be dismissed. In fact, in international relations, issues rarely die. They may only momentarily fade away to reappear at another historical juncture, and, perhaps, in a slightly different *avataar.*[1]

It is for this reason that India must not completely write off the CTBT from its nuclear calculations. While the issue at present may not overly preoccupy Indian nuclear diplomacy, nevertheless, a debate on its advantages and limitations must continue in order to arrive at a considered position on it. International circumstances have conspired to provide India with a rare opportunity where a discussion on the treaty can be conducted without any pressure—of time, or of the nature of decision that must be taken. This opportunity, then must be optimally utilised to objectively assess the treaty, its implications for India now that it is a state with nuclear weapons, and the options before the country. Such an exercise would prepare India for the moment, whenever that may be, when it might be called upon to take a decision on the treaty.

In 1996, when the treaty was wrapped up, rustled through the UN General Assembly instead of being passed through the CD where it had been negotiated, and opened for signature, India had decided that it would not subscribe to it for two reasons. First, because it did not establish any linkage with the issue of universal nuclear disarmament, a long standing Indian objective. India perceived that instead of being a step towards the eventual elimination of nuclear weapons, the treaty that emerged only appeared to drive nuclear testing into laboratories.

Modernisation of existing arsenals was likely to continue through computer modelling and simulation by those among the NWS who had such technological capability. For others not so equipped, their level of capability was being frozen at current levels. Overall, therefore, there appeared to be little willingness on the part of the NWS to *actually give up* their nuclear arsenals.

The second reason for India's non-acceptance of the CTBT was that it did not meet the country's security concerns. New Delhi formulated that the country's security was adversely affected by the continued presence of nuclear weapons worldwide and especially in India's neighbourhood. China was a nuclear weapon state recognised under the NPT. At the same time, Pakistan was also already known to possess nuclear weapons, acquired mainly through the transfer of tried and tested technologies from China. In the wake of a continuance of this flow of nuclear warheads and delivery systems, India could not have afforded to tie its hands with the CTBT. The need for an Indian nuclear deterrent was becoming more and more apparent. And, unless the CTBT was able to address the security issue in any comprehensive way, it was not considered to be in India's national interest to be a party to it.

The present need to reconsider this decision and reassesss the same treaty that had been consensually rejected in 1996 by the entire range of the political spectrum is triggered not by any alteration in the basics of the treaty, but because of a change in the frame of reference—at both the domestic and international levels. This chapter attempts to examine these changed parameters and to determine the extent to which they need to be taken into account before any decision on the CTBT is made. It then explores the various options before India on this issue, and the merits and demerits of each one of them. It concludes by advocating a choice that appears to be the most logical and sensible under the present set of international and domestic circumstances.

The Evolution of the CTBT

It is well documented that India had remained at the vanguard of the struggle for a comprehensive test ban ever since India's first Prime Minister Jawaharlal Nehru called for a halt to all forms of testing in 1954.[2] From then on, India consistently argued for a ban on nuclear tests as an important definitive and irreversible step towards the ultimate realisation of a nuclear weapon free world by putting a halt to the qualitative development, upgradation and improvement of nuclear weapons.[3] All through its presentation before the plenary of the CD, the main negotiating body that formally adopted the mandate to draft the test ban treaty in 1994,[4] Indian representatives pressed for the evolving CTBT to be placed in the context of total nuclear disarmament within a well-defined time-frame. India upheld that without such a linkage, steps like the CTBT or the proposed fissile material cut-off convention were "narrow and futile exercises aimed only at controlling non-nuclear weapon states, (and thereby) further strengthening the discrimination inherent in the non-proliferation regimes..."[5]

On the other hand, the US, that became keen to conclude a CTBT only in the early 1990s after having opposed it over the last four decades, saw it primarily as a non-proliferation tool. It perceived it as a means to prohibit other countries from testing in order to prevent both horizontal and vertical proliferation. On the one hand, the NNWS were bound to remain so since they could not indulge in nuclear testing. On the other hand, the NWS would be frozen at their current level of capability, thereby removing the possibility of a nuclear arms race.[6] Another underlying intention was also to ensnare India, and Pakistan into the non-proliferation regime through a route other than their acceptance of the NPT, since that appeared virtually impossible.

The difference in the Indian and US perceptions ultimately compelled India to disassociate itself from the treaty, despite having enthusiastically participated in the process of its drafting at the CD. In her statement at the plenary session of the CD on June 20, 1996, the Indian representative to the UN offices at Geneva stated, "The CTBT that we see emerging appears to be shaped more by the technological preferences of the nuclear weapon states rather than the imperatives of nuclear disarmament. This was not the CTBT that India envisaged in 1954. This cannot be the CTBT that India can be expected to accept".[7] It was also highlighted that India's national security was adversely affected by the prospect of some countries relying on nuclear weapons for their security while denying the same privilege to others.[8] Consequently, India refused to accept the CTBT since it was not able to "accept any restraints on its capability if other countries remain unwilling to accept the obligation to eliminate their nuclear weapons."[9]

Indian opposition to the treaty in the form in which it was emerging was further heightened with the inclusion of Article XIV. In an unprecedented move, this Article of the treaty specified preconditions for its activation that India found totally unacceptable. It introduced an ingenious entry-into-force clause that stipulated that to become operative, the treaty must acquire the ratification of all the 44 countries recognised by the IAEA as possessing and operating nuclear power or research reactors. And, this includes India. Clearly, the motivation for this provision was the intention to bring India into the CTBT fold at some point in the future. Expectedly though, it was objected to by India for being in violation of the Vienna Convention on the Law of Treaties. The strategy of denying a country the right of voluntary consent was accordingly condemned by the Indian representative, "...we cannot have our sovereign right not to sign the Treaty taken away and accept obligations on India that we cannot and will not accept."[10]

Indian non-acceptance of the CTBT made it impossible for the treaty to be presented as a consensus document by the CD to the UN General Assembly. The CD functions strictly on the principle of consensus and the certainty of an Indian veto was sought to be circumvented by having an identical draft presented to the UN General Assembly as a national paper by the Australian government with the co-sponsorship of 126 other countries. Though outraged at the complicity of some countries, even at this stage, India pressed for some sort of a linkage with nuclear disarmament. It proposed the inclusion of a paragraph in the preamble stating that the CTBT be seen as an "integral part of the commitment of the international community to achieve a complete elimination of all nuclear weapons within a time-bound framework." Unfortunately, however, the NWS were averse to any such commitment and the final treaty fell far short of Indian expectations. In the face of the intransigence of the NWS, a consensus had by then developed within India that it was not in the interest of the nation to subscribe to such a treaty. India, along with Bhutan and Libya, voted against the resolution, while five countries abstained. However, with 158 countries voting in favour of it, the resolution on the CTBT was adopted on September 10, 1996.

The UN secretary general, in his capacity as the depositary of the treaty, opened it for signature on September 24, 1996. Nearly six years down the line, it has been signed by over 166 states and ratified by 96. Out of the 44 specified nations,[11] all except India, Pakistan and North

Korea have signed it, but only 31 have ratified it yet. Out of these, France, the UK, and more recently Russia, are the three NWS to have done so. The US Senate has rejected the CTBT; China is yet to ratify it; India is debating options to arrive at a consensus; and Pakistan too remains undecided and reportedly engaged in a similar exercise.[12]

Need for Rebuilding Consensus

In 1996, when India had rejected the treaty, it had done so on the basis of a broad-based national consensus. In his *suo moto* statement to the Parliament, then External Affairs Minister I.K. Gujral alluded to this when he said, "The strength of the national consensus is reflected in the policy of this government and the course of action which was adopted during the discussions in New York."[13] However, the fact that this consensus today needs a rethink is prompted by a number of transformations that have since taken place in the country's nuclear policy, as well as in the surrounding international environment. These changes need to be adequately understood before any decision on the CTBT is taken.

A State With Nuclear Weapons

Most significantly, India's nuclear status today is different from what it was in 1996. From a country exercising restraint despite having demonstrated a nuclear capability in 1974, India is now a state with an overtly weaponised nuclear competence. In 1996, when it rejected the test ban, it was with the stated objective of not wanting to limit its future choices. With a regional security environment that was less than benign, and rather threatening to deteriorate with an increasing collusion between China and Pakistan on military and nuclear matters, India was averse to accepting a test ban without a definitive commitment on universal nuclear disarmament. The downslide in the regional scenario continued and finally led the government to decide in favour of conducting five nuclear weapon tests in May 1998.

Consequently, India is now a nuclear weapon state, even though the international community will not accept it as one under the NPT.[14] However, irrespective of whether it is granted the status or not, the fact remains that nuclear weapons now do need to be factored into the country's security strategy. Their primary objective, as stated in the draft nuclear doctrine put forth by the government for a countrywide debate, remains to provide an "effective, credible nuclear deterrence and adequate retaliatory capability should deterrence fail".[15] Nuclear deterrence, meanwhile, is itself a derivative of the kind of nuclear weapons, command and control structure and delivery capability that a country possesses. Therefore, the Indian decision on the CTBT must now take into account the kind of nuclear force structure it aspires to build and whether that would require any further testing. While scientists have argued both for and against further tests, the former Department of Atomic Energy (DAE) Chairman, Dr Chidambaram, during his tenure, had assured the country on more than one occasion that no further tests are required.[16] He had also expressed confidence in India's ability to conduct sub-critical tests. Similarly, Dr A.P.J. Kalam had stated in 1998 that the tests had "enhanced our design and simulation capability...We believe that subscribing to the CTBT will not affect our status as a nuclear weapon state."[17]

On the other hand, some eminent nuclear physicists such as Dr P.K. Iyengar have advocated further testing, especially if the Indian nuclear deterrent is to be based on compact, light and robust thermonuclear weapons that are truly tried and tested for reliability and safety.[18] These technical questions are crucial indeed and need to be adequately considered before India decides to join the CTBT. While India may have acquired the capability to develop a 200 KT (kilo tonne) weapon, there can be no denying that thermonuclear weapons do offer better weaponisation options since they require less fissile material, are lighter and hence more compact, and have better safety features. Having conducted only one test of a fusion device, can India be confident enough on this count? The decision on the CTBT must first of all rest on the technological parameters of India's nuclear deterrent and a healthy debate in the scientific community might be the best equipped to throw up answers to these questions.

Increasing Salience of Nuclear Weapons

The second major change in circumstances that should be brought to bear on any decision that India takes on the CTBT arises from an overall change in thinking on nuclear weapons. It may be recalled that the end of the Cold War had been accompanied with optimistic predictions on the likelihood of a willingness on the part of the two biggest NWS to renounce their nuclear arsenals in the wake of the end of the bipolar confrontation. The US-Russia relationship was beginning to be talked about in terms of partnership[19] and it was becoming fashionable to dismiss any future role for nuclear weapons. For instance, the Stimson Centre came out with a publication in the mid-1990s that argued that the character of international relations "is undergoing an irreversible transformation that would eventually invalidate rationales for weapons of mass destruction".[20] Some of this new thinking was propelled by initiatives such as those put forth by President Bush and reciprocated by President Mikhail Gorbachev in the early 1990s. These included hastening the process of elimination of strategic nuclear weapons covered under START I, putting an end to ground alert for nuclear armed bombers, halting the development of mobile land-based and new air-launched missiles, and ending further testing. The commencement of CTBT negotiations was then, a reflection of this new atmosphere. In fact, in 1995, at the time of the indefinite extension of the NPT, the NWS promised to push for an early conclusion of the CTBT, "no later than 1996".[21] The negotiations did manage to conclude a year later, though the manner in which this was done and the form in which the treaty evolved, leaving intact the ability of the NWS to conduct other forms of non-explosive testing, meant that it would have little scope of ever contributing to the realisation of a nuclear weapon-free world.

Since 1996, when the CTBT was first opened for signature, with President Clinton being the first one to sign it, several developments demonstrate a trend away from undertaking any worthwhile commitment on nuclear disarmament. Rather, developments in each one of the NWS point towards an increasing salience of nuclear weapons in their national defence strategies. The US itself appears to have led the way with several actions and overtures that have exhibited its desire to keep nuclear deterrence at the heart of its national security strategy. In fact, less than a year after signing the CTBT, the US had reaffirmed that "strategic nuclear deterrence remains a key US military priority."[22] Recent planning documents of the US Joint Chiefs of Staff even contemplate the use of nuclear weapons against the use of chemical and biological weapons. By expanding the core mission of nuclear weapons from merely nuclear deterrence to providing a hedge against other means of mass destruction, such a move adds value to nuclear weapons and could cause further nuclear proliferation. Nevertheless, the US continues to regularly earmark huge amounts of money for updating and maintaining its formidable arsenal. An elaborate and expensive Stockpile Stewardship and Management Programme (SSMP) has also been put in place to ensure the refinement and reliability of the existing stockpile. It is specifically entrusted with the task of "guarantee(ing) the safety and reliability of our nuclear weapons under the CTBT." And yet, despite all this the US Senate did not deem it fit to ratify the treaty.

In fact, a Republican senator, Richard Lugar, in his statement explaining his vote against the treaty, quoted a nuclear scientist to suggest that the challenge of maintaining the viability of a nuclear stockpile without testing is like "walking an obstacle course in the dark when your last glimpse of light was a flash of lightning back in 1992."[23] Most of the other Republicans too subscribed to this viewpoint and voted to defeat the treaty since "it jeopardised our nation's (US) security by undermining the US nuclear deterrent that has served our country so well for the past 50 years".[24] If a country like the US that has archived data from more than 1,000 nuclear tests is so diffident on binding its hands with the CTBT, then what does this imply for a country like India? In fact, it would be worthwhile to consider the two main arguments that the US senators made while voting against the treaty in October 1999:

- Testing is imperative if the country is to maintain a safe and reliable nuclear weapons arsenal. In the long run, the CTBT would jeopardise the ability of the country to maintain the safety and reliability of the nuclear arsenal. James Robinson, director of Sandia National Laboratory, stated as much when he said, "To forgo testing is to live with uncertainty."[25]
- The treaty is basically unenforceable and unverifiable and, hence, would have little impact on the problem of proliferation. Rather, it could propel proliferation if countries become unsure of the US nuclear umbrella.

Following the Senate rejection of the CTBT in 1999, then US President Bill Clinton asked General John Shalikashvili to prepare a report that would address the concerns of the Republicans. After a ten-month review, the general compiled a report that strongly recommended a bipartisan approval to the CTBT. However, nothing came out of the exercise. Rather US contempt for the CTBT attains even more sinister proportions when seen against the American insistence on deploying an NMD in violation of the ABM Treaty of 1972. With the demise of the ABM Treaty, there is an air of resignation around the world on the inevitability of the NMD. This, however, barely masks a sense of frustration and discontentment with the state of affairs. The NWS such as China and France are on the defensive and have criticised the US endeavours as an unacceptable effort designed to achieve strategic superiority at the cost of disrupting global and regional strategic balance and stability. The Chinese representative, Ambassador Sha Zukang, made it clear at the NPT RevCon in May 2000 that NMD deployment "is tantamount to a nuclear arms build-up, which will not only bring severe damage to...the global strategic balance and stability, but also...impede the international disarmament process, and thus shatter the prerequisite and basis for international nuclear non-proliferation."[26] An ominous statement, indeed and with implications for India. A US with missile defence would compel China to reassess and upgrade its own nuclear capability and this would, in turn, impinge upon India's security.

Further questions have been raised in the context of the CTBT with the disclosure/leakage in early 2002 of the latest Nuclear Posture Review (NPR) conducted by the Pentagon at the behest of Congress. Two aspects in this regard need particular attention. The first of these is more directly related to the CTBT as it reiterates opposition to the American ratification of the CTBT, and at the same time recommends that the country should reduce the time necessary to prepare its nuclear test sites from the current period of two years. This is despite the fact that the US SSMP is a virtual nuclear testing facility. The SSMP ensures the continued viability of the US nuclear weapon infrastructure, trains a new generation of weapon scientists, and takes up refinement of weapon designs and production processes. In fact, weapon scientists have suggested that the programme has the capability to meet emerging military requirements through new weapons that "could be deployed...without the need for nuclear tests."[27] Yet, the NPR proposes accelerating test site readiness, primarily to aid US plans of possibly modifying its existing nuclear weapons to give them "greater capability against...hard targets and deeply buried targets."[28]

As of now, a 1994 legislation prohibits laboratories from undertaking research and development of new weapon designs such as the small, high precision and bunker buster-types that are being talked about, but Press reports have suggested that the issue could figure in the coming sessions of Congress. It is well known that weapon labs, as also some politicians and military officials have long pushed for low yield weapons that could be more usable, especially to defeat hardened and deeply buried targets such as bunkers housing terrorists, or biological and chemical weapons facilities. The US military experience with Iraq over the last decade, and its more recent battle against the Taliban who used caves as their hideouts can only strengthen the case for these mini-nukes. A report published by the National Institute for Public Policy in January 2001 also supports such weapons for their "unique targeting capabilities," especially against chemical and biological weapons in states of concern. The same case has been made by the director of the Sandia Laboratories, "...the yields of weapons left over from the Cold War are too high for addressing the deterrence requirements of a multipolar, widely proliferated world. Without rectifying that situation, we would end up being self-deterred."[29]

Any US effort, however, to scrap the CTBT even in practice through the development of newer weapon designs is fraught with danger. It shall severely strain the test moratorium in place worldwide, dim hopes for the eventual entry into force of the CTBT, and push other countries to factor nuclear tests in developing or modifying their nuclear weapons by placing a question mark over US intentions in the matter. It may be recalled that in the year 2001, President Bush had even suggested that the US would understand if the Chinese were to undertake further nuclear tests in the face of an NMD deployment. The conduct of nuclear tests by any country at this historical juncture, however, could open a Pandora's box.

While President Bush has pledged to uphold his country's moratorium on testing, yet jettisoning the treaty leaves open options for future testing and modernisation of weapons.[30]

Implications of the Verification Mechanism

A third factor that must be considered before any decision on the CTBT is taken relates to the treaty's verification and enforcement mechanism that India will have to contend with if the treaty ever becomes operational. Even though the likelihood of the treaty entering into force anywhere in the near future remains remote, it would still do India well to understand the mechanics of the verification regime that it envisages. The CTBT shall be implemented through vast machinery involved in the International Monitoring System (IMS) to detect and identify nuclear explosions, an International Data Centre (IDC) and other inspection measures. The IMS, when fully functional, would comprise a global network of 321 monitoring stations capable of detecting nuclear explosions around the world. The network would include 50 primary and 120 auxiliary seismic stations capable of detecting explosions of even a low yield and being able to distinguish such explosions from earthquakes. It would also include 80 radio nuclide stations to identify radioactive particles released during an explosion and 16 laboratories for sample analysis to support these stations. Further, 60 infrasound stations and 11 hydroacoustic stations are to be installed to detect acoustic signals in the atmosphere or under water that might be generated by an explosion.[31]

All these stations would be connected online to an IDC in Vienna from where states parties could receive both raw and processed information. The IDC represents a significant technological challenge since phenomenal amounts of data from different technologies would be continuously pouring into the centre from stations all over the world, to be consolidated into a single data analysis system at a very rapid pace. During the CTBT negotiations, the developing countries, which do not possess the capability to analyse data from the IDC, wanted to have the IDC screen the data in order to render expert opinion on ambiguous events. However, the technologically advanced countries felt that such interpretations should be the responsibility of the states parties.

Expectedly, the mechanism planned shall not be a cheap one. The verification system is estimated to cost $150 million for equipment and installation, and to have annual operating and maintenance costs of around $85 million when fully operational. This financial burden shall have to be shouldered by all countries party to the treaty.

Notwithstanding such elaborate procedures, and their cost, it remains a fact that aspersions have been cast on the level of efficacy that the verification mechanism would achieve. For instance, advances in mining technologies have enabled the smothering of nuclear tests without their being detected. Tests could be masked or muffled in mines, underground cavities, salt domes, or other geological formations. In any case, it is a known fact that the IMS can detect only explosive tests above a level of one KT, and that too in a non-clandestine environment. This might not always be so and it could be possible for some countries to conduct tests of a low yield without their being detected. In fact, some analysts have questioned the efficacy of the monitoring mechanism, given that initial US announcements on Pokhran II dismissed it as an earthquake or a test of much lower yield than the Indian scientists have claimed!

Another problem may arise from the rather intrusive inspection regime envisaged under the treaty. On-site inspections, to be held with 30 votes out of a 51-strong Executive Council, could present themselves as tools of harassment by some states against others. At the same time, it may be noted that timelines imposed by the treaty for on-site inspections permit considerable scope for cover-up or deterioration of evidence. The country that is to be inspected could also choose countries from where the inspectors would arrive and block them from countries that it does not trust. Provisions such as these are inevitably amenable to manipulation and abuse and India must be aware of these risks.

India's Options

It is against the backdrop of these three factors that this chapter attempts to explore India's options and the merits and demerits of each.

Decision One: Sign and Ratify

One clear option presents itself in signing and ratifying the CTBT. After Pokhran II, India now has a proven and demonstrated capability of producing at least 200 KT thermonuclear warheads—not a state of the art capability, though it is not a small yield either.[32] With an effective delivery capability, it should be enough for the purpose of deterrence. Meanwhile, in any case, the government has declared a unilateral moratorium on further testing and does not appear to be in any mood to conduct any more tests immediately. So, several nuclear strategists have argued that India need not hesitate in signing and ratifying the treaty. Especially, since in any case, if at any future date, a need is felt to carry out more tests, India could always withdraw from the treaty on the grounds of national security.

The option, however, needs to be scrutinised a little more carefully, particularly in terms of its advantages and disadvantages. As regards the former, it may be said that by accepting the treaty, India would definitely win international applause and goodwill. Also, it would gain the upper moral hand vis-à-vis the US and other countries holding back on ratification. As one Indian analyst has put it, "Signing the CTBT and giving the US (and the rest of the world community) some of the reassurances on our nuclear weapons programme will put the seal of international acceptance on India's emergence as a mature and stable nation-state, destined for a place in the first tier of nations."[33]

Another possible consequence could be that greater pressure could be brought to bear on China, and Pakistan, thereby stopping the evolution of more advanced weapons in the neighbourhood. However, this remains only a possibility and it may turn out that despite Indian endorsement of the treaty, the other two hold back and continue to improve their nuclear arsenals. This would have serious implications for Indian security.

One more benefit, though, can be visualised. After signing and ratifying the treaty, India would enjoy the privilege of participating in the special conferences that would be convened every year until the treaty enters into force. As has been mentioned earlier, Article XIV of the CTBT stipulates that it could enter into force only after all the 44 countries have deposited their instruments of ratification with the UN secretary general. However, anticipating that not all countries might ratify the treaty promptly, paragraph 2 of the same Article provides that states that have already ratified the treaty could call upon the secretary general to convene a special conference three years after its being opened for signature. Such a conference, then, would be held each year until the treaty enters into force. States parties that have already ratified the treaty would have the right to sit in at the conference, while signatory states that have not ratified would attend the conference only as observers with no role in decision-making.

The purpose of the conference would be to "consider and decide what measures consistent with international law may be undertaken to accelerate the ratification process." In this forum, then, India, as a state party, could join forces with other traditional supporters of the treaty such as Japan, UK, France, Netherlands, South Africa, Australia, New Zealand and Mexico in trying to make the treaty more favourable to an eventual elimination of nuclear weapons. Most of these countries have articulated their positions in favour of nuclear disarmament at various fora. Hence, it might be worthwhile for India to wrest the initiative to push its case for amending the treaty to make it more conducive to its own security interests.

However, it should also be understood that there are limits to what India, or any other state party can do at the special conference. In fact, these conferences have been regularly held since 1999, but have achieved little by way of pushing the cause of the CTBT. They have rather been constrained by several factors. For one, Article XIV does not empower the conference to waive the entry into force requirements. Nor can it amend the treaty to change this requirement. Hence, the conference can mainly dedicate its efforts toward creating conditions of political pressure to bringing about the desired ratification. This could be done by strengthening the international consensus against testing and building pressure for ratification by the bigger states. Secondly, the conference can set up mechanisms under which the CTBT and its verification system could be provisionally applied. But this would have its own problems in terms of decision-making and authority, payment sharing, and differentiating the role and responsibilities of countries that have signed but not ratified and those that have ratified. Therefore, even after ratifying, India would not essentially have a free hand in any way remoulding the CTBT to suit its own objectives. Though, of course, it shall be able to at least sit in at the negotiations and keep a watch on the proceedings.

Apparently, therefore, the only tangible benefit of exercising this option would be to gain international goodwill. But should this be deemed enough? Having waited it out for 25 years before exercising an option in favour of weaponisation and having borne the brunt of international criticism and sanctions after having finally tested in 1998, should regaining international goodwill be the only reason for signing the CTBT? Should this be enough to make India go back on its vow that it would not sign the CTBT in its present form, "not now, not later"? Especially since the CTBT is still the same—not comprehensive since it bans only nuclear explosions while allowing other higher forms of testing; and not a test ban since it allows non-explosive sub-critical tests. Advanced computer simulation using extensive data relating to previous explosive testing and weapon related applications of laser ignition could lead to a fourth generation of nuclear weapons, even with a ban on explosive testing. And this seems highly probable given the international trend away from nuclear disarmament.

Neither has the regional security scenario changed in any favourable way. Rather, both China and Pakistan continue with their military and nuclear programmes, and if anything there is evidence of a close nexus between the two on both these issues. Given these conditions, should earning goodwill be sufficient justification for signing and ratifying the treaty? And, would the world community be ready to include India in the "first tier of nations" merely on the basis of its signature on the CTBT? It appears highly unlikely that they would not press for far greater concessions from India such as compelling it to accept the NPT too.

Moreover, withdrawal from the treaty, if ever that should become necessary, would entail its own problems of inviting severe international ramifications. Also, after withdrawal, it might be unrealistic to expect quick testing in an emergency. Lab directors in the US have noted that it would take the US at least two to three years of preparation before a test can be conducted since the testing infrastructure would have deteriorated, if not disappeared, over time. The time taken by India could only be more. A withdrawal would also send out a signal that the efficacy of the Indian nuclear deterrent had deteriorated. The ensuing reduction in confidence in the reliability of the nuclear stockpile could be highly demoralising and destabilising for the country, both internally and externally.

Decision Two: Sign, Yes. Ratify, No

Another option that has been advocated as regards the CTBT is that India could sign the treaty but reserve the right to ratify it for some later date. Such a step could help India gain some amount of moral edge, while still retaining a part of the leverage for the future. While this may superficially seem like a sensible option, however, it should be understood that international pressure would not stay off for too long. Nor can it be expected to be satisfied with the mere signature. Rather, pressure to ratify could mount even further. As has been pointed out by two former Indian representatives to the CD, "Once India signs the CTBT, the American pressure will shift to ratification, to a moratorium on production of fissile material and eventually to signing the NPT. The nuclear goalpost for India will remain a moving one..."[34]

Moreover, it need also be said that though as a signatory that has not ratified the treaty, India may not be obligated to abide by each and every term of the agreement, yet, under customary international law, it shall still be obliged to refrain from actions that go against the "object and purpose" of the treaty. Therefore, it shall in any case be obliged to refrain from conducting tests until it makes clear its intention not to ratify the treaty at all.

Moreover, while signing and not ratifying the treaty might have been an option for the US given its presidential system of governance, the same does not hold for India. In our parliamentary system, it is the prime minister who has to take the decision and it is not mandatory for the treaty to be subjected to any kind of scrutiny by Parliament. In fact, the Indian Constitution does not have any provision for ratification of treaties by Parliament.[35] Treaty making, in the Indian system, is entirely an executive act performed by the prime minister and then presented to the president for his nod that can hardly not be forthcoming once the prime minister recommends it to him. Therefore, holding back on ratification would leave nothing more than a token leverage in India's hand without bringing forth any tangible benefits.

Decision Three: Sitting on the Fence

As a third option, India could continue to sit on the fence as it does now. Would this bring any benefits? For one, the option of nuclear testing would remain open. While the significance of this may be reduced by the fact that the treaty has a withdrawal clause, yet it may be said that withdrawal in practice is not easy to exercise, given the international opprobrium that would surely follow and the dangerous precedent that it would set. Therefore, one response could well question the need to get into that uncomfortable position at all. One could continue to stay out of the treaty and observe how future developments unfold.

However, the disadvantage of this option would be that even as India shirks taking a real decision, it would not be testing either and not gain anything from the moratorium since its intentions would still be suspect. Moreover, India would then also be resigning itself to only responding to the developments instead of proactively pushing for things to move as it desires—at the internal level, in terms of gaining some incentives, and at the international level, in terms of shaping the global agenda to suit its security and other concerns.

Decision Four: Seek Incentives for Signing

A fourth option presents itself as being the most viable one in the present circumstances. It may also be seen as mixing a bit of idealism with true blue pragmatism and *realpolitik,* the buzzwords of modern times. India could make it clear that it could consider subscribing to the treaty, provided a few of its conditions or concerns are adequately met. However, to carry through such an option would require utmost political and diplomatic skill. A clear message would have to be conveyed that in accepting the CTBT in its present form, India is compromising its earlier objections to the same and, therefore, they should be addressed in another form or mode. At the same time, India would have to press the point that by accepting the CTBT even after having carried out the tests in 1998, it would be tying its hands on the kind of nuclear deterrent it could develop (on the weight-to-yield ratio of its nuclear warheads) and for this it should be adequately compensated by addressing its security concerns in another form or forum.

In fact, ever since the nuclear tests in May 1998, Indian officials have repeatedly pointed out that ways and means need to be worked out to enable India to sign the treaty. One of these would involve sorting out the problem of the treaty's linkage (or the lack of it) to nuclear disarmament. However, realising that amendments to the treaty were not going to be easy, Prime Minister Vajpayee had even suggested on September 16, 1998, "I know the Treaty is not going to be amended, but certain other arrangements could be made which will facilitate our work...particularly regarding the transfer of high technology."[36]

This proposition, in turn, gives rise to three questions: first, what could or should India demand in return for its acceptance of the CTBT? Second, why and whether the international community would or could accept India's demands? And third, just in case India does get a good bargain for its price, then would it be the right move to go ahead and sign?

The Demands

As far as the first point is concerned, there are a few proposals that India could submit for consideration to the major powers in return for its signature. The most important of these could perhaps be recognition of its nuclear weapon status. This would be important for several reasons. For one, it would legitimise India's right to conduct sub-critical tests, a right that only the NWS have retained for themselves. In parallel negotiations among the P-5, even as the CD was negotiating the treaty, the NWS had defined for themselves a set of related activities that they could engage in despite the CTBT. These include high-tech underground tests at sub-critical level, simulated nuclear explosions in labs and the sharing of data and technologies among themselves. Therefore, unless India is recognised as a nuclear weapon state, it would not be legally permitted to undertake these activities.

Secondly, India could seek a lifting of the export controls on high-end technologies, especially on nuclear and space-based applications for peaceful use. Once India is recognised as a nuclear weapon state, the transfer of such technology, especially that of nuclear power reactors, could become easier and without any insistence on full scope safeguards. And, thirdly, India could seek some sort of a movement towards universal nuclear disarmament, a national objective as per the draft nuclear doctrine.

Would the Demands be Met?

While these demands could well be formulated, the more important question is, why should the international community (read major powers) agree to pay such a price for securing Indian acceptance of the CTBT? In fact, would they agree at all? Perhaps not. But then there is nothing lost by pitching high. And there is a slight chance that some of these demands might be accepted. Of course, India would need to back these demands by pressing on and constantly highlighting the advantages that will accrue to the international community and the non-proliferation regime through India's signature. For instance, the Indian endorsement of the CTBT at this juncture when its state appears so uncertain could set in motion a positive trend. In fact, in 1998, when India went nuclear, it was accused of heightening the probability of other countries following suit. By the same token, it may be argued that if it were now to sign the treaty, would it not set in motion a trend in favour of the CTBT? For other international players like the EU countries, Canada and Japan, the Indian signature could smoothen their entry into a country with a huge market with massive potential. They are likely to come in anyway, irrespective of the Indian signature, but India's signature would provide them with a face saving device. At the same time, several disenchanted non-nuclear parties of the NPT could be won over by emphasising that by linking Indian acceptance of the CTBT to some sort of a movement on nuclear disarmament, India was in fact championing their cause. Concerted pressure then may be brought to bear on the NWS and their allies.

Of course, such actions would call for the skilfull use of Indian diplomacy at various levels. It would also take some time for any results to show up. But then, what is the hurry? Irrespective of whether the Indian government signs the CTBT now or not, it is likely to remain in a state of limbo till the US Administration decides otherwise. In the meanwhile, the issue of NMD deployment would have elicited a much clearer and concrete response from China. If China were to conclude that its nuclear deterrent is unable to match up to the US arsenal as being fortified by the SSMP and an eventual NMD, then it might decide in favour of resuming testing. Naturally, then, India too would have to reconsider the state of its own nuclear deterrent and this would impact on any decision on further testing. Therefore, while the international and regional scenario remains so uncertain, would it not suit India's interests not to be hasty and instead seek worthwhile incentives for doing so?

Strike a Deal?

At the same time, it should also be considered that just in case India does receive a good bargain for its signature, then would there be any harm in going ahead? Perhaps not, for two reasons. First, the CTBT would allow India to retain its nuclear deterrent. The treaty does not require its subscribers to dismantle their existing stockpiles. Nor does it prevent them from maintaining those stockpiles through scientific means. Second, the treaty will help to create a worldwide nuclear status quo. Parties to the CTBT would be able to conduct nuclear explosive tests to improve the existing weapons or develop stronger ones. Of course, this would mean that the Indian nuclear deterrent would always remain stronger vis-à-vis Pakistan but weaker against China. However, it must be understood that a nuclear deterrent is not the direct function of only the yield or number of nuclear weapons. Rather, it is the sum total of other components as well, such as the range and reliability of delivery mechanisms, the robustness of the command and control systems and the perception in the enemy's mind about the survivability of the arsenal. Therefore, the relative Chinese nuclear superiority could be met with improvements in the other components of nuclear deterrence.

Above all, from the Indian point of view, the verification mechanisms and their enforceability would provide valuable lessons for the eventual implementation of a complete elimination of nuclear weapons. It could prove or disprove whether such treaties can at all be verifiable.

Conclusion

In conclusion, it may be stated that at present the CTBT lies in a state of dormancy. To resuscitate it actually lies in the hands of the US. However, it would have to do so not only by ratifying the treaty, but also by articulating a new approach to arms control and multilateral treaties, since Russia and China are likely to hinge their approach to the treaty and further arms control or non-proliferation measures on the US approach to both. Meanwhile, the US seems to be caught in a dilemma on both counts. On the CTBT, a treaty once rejected in the Senate cannot be resubmitted for consideration without amendments. No amendments to the treaty are, however, possible unless it first comes into force, But, it cannot come into force unless the 44 specified countries ratify it and the US is one of them. On the issue of the NMD, its deployment has the potential to unhinge all arms control regimes, contribute to the initiation of a nuclear arms race and even further nuclear testing. Yet, the issue has acquired a momentum of its own, given the vested interests involved—of the bureaucracy, the military and the private industry.

Given such a scenario, it would be best for India not to commit itself to the CTBT unless it can get something really worthwhile in return. In fact, at the moment it would be imperative to keep a close watch on the evolving international developments on the issues discussed above. The fate of the CTBT, for the time being, remains tied to the generally discouraging trend in arms control and nuclear non-proliferation.

Notes

1. The reappearance of the Strategic Defence Initiative or the Star Wars in the modern day version of the NMD is a case in point.
2. Jawaharlal Nehru, "Standstill Agreement," in *India and Disarmament: An Anthology* (New Delhi: Ministry of External Affairs, Government of India, 1988), p. 33.
3. In 1963, when the Partial Test Ban Treaty (PTBT) was concluded, India subscribed to it in the belief that it would lead to a comprehensive ban and then to disarmament. At the Eighteen Nation Disarmament Committee (ENDC), the Indian representative pointed out that the importance of the treaty "does not lie so much in what the treaty actually says, as in what it means and the hopes it arouses," UNDC/PV 156, August 29, 1963, p. 13 as quoted in K.K. Pathak, *Nuclear Policy of India* (New Delhi: Gitanjali Prakashan, 1980), p. 86.
4. It may be mentioned that the CD had long been involved with the issue of a test ban. A subsidiary body on the item had been established way back in 1982, but disagreement over the mandate had blocked all progress.
5. Statement made by Arundhati Ghose, ambassador/permanent representative of India to UN Offices at Geneva, in the Plenary of the CD on September 21, 1995. Cited in *Statements by India on CTBT, 1993-1996* (New Delhi: Ministry of External Affairs, Government of India, 1997).

6. See Glen T. Seaborg and Benjamin S. Loeb, *Kennedy, Khruschev and the Test Ban* (California: University of California Press, 1981) pp. 165-166.

7. Statement made by Arundhati Ghose, ambassador/permanent representative of India to UN Offices at Geneva, in the Plenary of the CD on June 20, 1996. See n. 5.

8. Ironically enough, it is the US that has now quoted national security as the reason for denying ratification to the CTBT. One of the Republican senators in his argument against the treaty demanded "real arms control—not the show, not the empty symbols, not the flimflam treaties that cannot accomplish their purposes." Another Republican senator described the treaty as simply banning nuclear explosions of any yield exceeding zero. "It is not a treaty by which states which currently have nuclear weapons agree to give them up, reduce their numbers, even stop their development or agree not to give them to others." Does this not vindicate the Indian stance that the CTBT is not a true non-proliferation and definitely not a disarmament measure?

9. See n. 5.

10. Statement made by Arundhati Ghose, ambassador/permanent representative of India to UN Offices at Geneva, in the Plenary of the CD on August 8, 1996. Ibid.

11. The 44 states are Algeria, Argentina, Australia, Austria, Bangladesh, Belgium, Brazil, Bulgaria, Canada, Chile, China, Colombia, Democratic People's Republic of Korea, Egypt, Finland, France, Germany, Hungary, India, Indonesia, Iran, Israel, Italy, Japan, Mexico, Netherlands, Norway, Pakistan, Peru, Poland, Romania, Republic of Korea, Russian Federation, Slovakia, South Africa, Spain, Sweden, Switzerland, Turkey, Ukraine, UK, USA, Vietnam, Zaire.

12. In March 2001, reports appeared in the Pakistani and international media on Islamabad's intention to sign the treaty and thereby delink it from the Indian position on the same. However, no official announcement to this effect was forthcoming and one cannot dismiss as mere coincidence the emergence of such statements whenever an instalment of some sort of aid package is due for the country. Naturally, Pakistan sees this as a useful bargaining chip and would want to make the most of it.

13. Suo Moto Statement made by I.K. Gujral, minister of external affairs, in the Parliament, regarding the CTBT on September 11, 1996.

14. Under the NPT, a nuclear weapon state is one which has manufactured and exploded a nuclear weapon or other nuclear explosive device before January 1, 1967.

15. For text of the draft nuclear doctrine put forth by the National Security Advisory Board, see *Strategic Digest,* (IDSA) September 1999.

16. Interview conducted by Nirmala George with Dr. R. Chidambaram, http://www.meadev.gov.in

17. "Kalam Proposes, Fernandes Opposes," *The Times of India,* September 22, 1998.

18. P.K. Iyengar, "In Testing Times: Repercussions of Signing the CTBT," *The Times of India,* February 17, 2000.
19. Several Russian strategic thinkers and even government officials of the time have been quoted as saying that Russia was keen to forge a partnership with the West. Some even ventured to suggest an eventual Russian integration with the West.
20. This report prepared by Barry Blechman and Cathleen Fisher is mentioned in Robert Manning, "The Nuclear Age: The Next Chapter," *Foreign Policy,* Winter 1997-98, p. 76.
21. The NPT Review and Extension Conference adopted a decision entitled "Principles and Objectives for Nuclear Non-Proliferation and Disarmament" that contained a programme of action for the full realisation and effective implementation of its Article VI that calls for efforts towards nuclear disarmament. The CTBT was envisaged as one of the steps in this direction.
22. Annual Report of the US Department of Defence as quoted in Office of the Secretary of Defence, *Nuclear Weapons System Sustainment Programmes* (Washington DC: Department of Defence, May 1997).
23. Richard Lugar, "Better Safe…", *Bulletin of Atomic Scientists,* January-February 2000.
24. Senator Kyl speaking on the CTBT vote as quoted in Senate Congressional Record, 106th Congress, 1st Session, Document no. 145 Cong Rec S 12690.
25. As quoted by Senator Warner while arguing against the CTBT during the Senate debate. See Congressional Records, 106th Congress, 1st Session, Document no. 145 Cong Rec S 12690.
26. "NMD Deployment: Creating Global Instability", BASIC Press release, July 6, 2000, http://www.basicint.org
27. Director, Sandia Labs, as quoted in "US Nuclear Weapons Policies, BMD, and the Quest for Weapons in Space: Military R&D and the New Arms Race," *Western States Legal Foundation Information Bulletin,* Summer 2000.
28. BASIC, "NPR Point-Counterpoint", http://www basicint.org
29. As quoted in Robert W. Nelson, "Low-Yield Earth-Penetrating Nuclear Weapons," *FAS' Public Interest Report,* January-February 2001. http://www.fas.org.
30. Theresa Hitchens, "The US Nuclear Debate: Issues of Concern," BASIC Paper no. 35, February 2001.
31. Sandra Glass, "The CTBT: Two Years On," *Disarmament Diplomacy,* issue no. 29, 1998.
32. The yield of the bomb dropped on Hiroshima was 15 KT and the devastation was of a horrendous magnitude.
33. Prem Shankar Jha, "Significance of the CTBT," *The Hindu,* January 10, 2000.

34. Prakash Shah and Arundhati Ghose, "Signing the CTBT: Implications for India's Security," *The Times of India,* January 25, 2000.

35. For greater details on a constitutional point of view on this subject, see Subhash C. Kashyap, "On Tricky Ground: Route to Signing the CTBT," *The Times of India,* April 22, 2000.

36. Sandra Glass, "CTBT Update," *Disarmament Diplomacy,* issue no. 29, 1998.

5

FMCT: The Last of Nuclear Arms Control?

Manpreet Sethi

Amongst the efforts currently underway in the field of arms control, the negotiations on the FMCT deserve a special mention since the conclusion of this treaty was identified as the next step that the international community would focus on after the conclusion of the CTBT in 1996. The CD, the main body undertaking the process, has since then convened several sessions every year to thrash out a treaty that would promote non-proliferation through placing a cap on further production of fissile material. Nonetheless, it has little to show for all the time spent at these sessions.

The reasons for this, of course, are many. For one, the treaty covers a lot of complex issues and traverses ground that has not been gone over earlier. Therefore, negotiations have to be slow to accommodate different country positions. Secondly, and more importantly though, nuclear arms control has become a casualty of a host of new international developments. It no longer occupies the pride of place in the contemporary international landscape that it did merely a decade ago and rather is increasingly and ever more ostensibly being used as a bargaining chip in bilateral and multilateral negotiations. Besides, the US, that has always managed to set the agenda on arms control, appears caught up in redefining its threat perceptions in the wake of September 11, waging a war against terrorism and adopting a unilateralist stance. With a predominance of its "go-it-alone" approach, multilateral arms control processes remain adrift. In fact, the FMCT negotiations have become mired in a number of other issues and the treaty is nowhere near a quick conclusion.

Why have things come to such a pass? What are the issues that now agitate international relations to the detriment of arms control measures? This chapter focusses particularly on the FMCT. It seeks to understand what the treaty would signify and explores the issues that are currently exercising the minds of the negotiators. The chapter concludes by spelling out all the implications of this for India. It flags the issues that India must especially deliberate upon with regard to the FMCT.

The FMCT

The issue of proscribing fissile material production has been on the international agenda since at least 1993. The UNGA had then adopted a resolution calling for the negotiation of a non-discriminatory, multilateral, and effectively verifiable treaty.[1] In 1994, the CD had agreed that it was the appropriate forum for the negotiation of such a treaty. One year later, the CD agreed to establish a mandate for an ad hoc committee. The mandate, after much debate, took shape in the form of the Shannon Report, named after the then Canadian ambassador who led the CD on the issue.[2] In 1998, the CD agreed to convene an ad hoc committee for negotiating the fissban. The committee, however, has failed because of the inability of the countries to agree on a complete programme of work. Consequently, the treaty is nowhere near any conclusive state of being.

Simply put, a treaty on fissile material cut-off encapsulates a proposal for the prohibition on the production of fissile material for nuclear weapons. An FMCT would, hence, imply a set of three commitments on the part of its member states:

(a) To desist from manufacturing highly enriched uranium (HEU) or separating plutonium for either the construction of nuclear weapons or research and development in that field.

(b) To refrain from assisting other states in this regard.

(c) To accept IAEA safeguards for verification of the implementation of these commitments.

When in place, the FMCT, would achieve multiple objectives. One, it would limit the size of the potential nuclear arsenals because fissile material constitutes the fundamental ingredient of any nuclear weapon. The effect of the ban would be even more stifling if existing stockpiles are also brought under the treaty purview. Second, it would obviate the chance of countries going back on their reductions, especially if the fissile material is transferred from dismantled weapons and other unsafeguarded stockpiles to non-weapons use or disposal under international safeguards. Third, it would strengthen the non-proliferation regime by opening nuclear facilities in all states to international inspections.

In fact, in this regard, the countries that have nuclear weapon capabilities but are outside the ambit of the NPT (India, Pakistan and Israel) have perceived the FMCT as an attempt at pulling them into accepting full-scope international safeguards. Their perception was echoed in a RAND Report published in 1995. It stated that the treaty "is aimed particularly at India, Pakistan and Israel, which are undeclared nuclear weapon states and are unlikely to join the NPT."[3] The only states capable of, and legally not encumbered from, producing fissile material for weapons use are the NWS within the NPT and the three states with nuclear capabilities outside the NPT. Of these eight countries, Britain, France, Russia and the United States, have anyway declared unilateral moratoria on further production of fissile material. China too, claims to have ceased production, although there is no official evidence to this effect. Therefore, the main merit of a fissban treaty lies in bringing the non-NPT nuclear states under international safeguards. Not surprisingly, then, the three countries have been wary of the treaty since it could mean opening their nuclear facilities to international monitoring mechanisms.

However, within the first few rounds of negotiations on the treaty, and in fact even when the mandate for the ad hoc committee was being decided, several other contentious issues had surfaced to divide the international community. These continue to weigh down the negotiations and the FMCT remains elusive.

Squabbles Over Scope

The most important of these relate to the delineation of the scope of the treaty—whether it should be confined only to future production of fissile material, or include existing stockpiles of fissile material also within its purview. Conflicting viewpoints hold up the conclusion of a mandate for negotiations on the FMCT.[4]

The five NWS (as also India) favour a "look ahead" orientation for the treaty by limiting it to verifying only a future cut-off in fissile material production. However, contesting this viewpoint are some other countries such as Egypt, Pakistan, Iran and Algeria that support the inclusion of provisions within the treaty for reductions in existing stockpiles too. In fact, on June 9, 1994, the Pakistani representative at the CD had sought a schedule for the progressive transfer of those stocks to safeguards, so that unsafeguarded stocks could be "equalised" at the lowest possible level.[5] In 1998, Pakistan reiterated, "We cannot agree to freeze inequality."[6]

In an effort to break the stalemate, Ambassador Shannon in his report of March 24, 1995, conceded that the "mandate for the establishment of the ad hoc committee does not preclude any delegation from raising for consideration in the ad hoc committee any of the above mentioned issues."[7] This was interpreted as keeping the focus on a future cut-off while allowing for discussion on the management of existing stockpiles as well. Hence, what was agreed to in 1995 was a broad and rather ambiguous mandate that excluded no viewpoint and only aimed to paper over the differences to put the negotiations back on track.

However, it soon became clear that despite clearing the first hurdle, the road ahead was going to be long and tortuous. Negotiations have since then remained stalled over two other issues. The first pertains to the position that the treaty should occupy in the larger scheme of universal nuclear disarmament. Most of the G-21 non-aligned states perceive the FMCT as an element of a larger programme of action for the ultimate elimination of nuclear weapons. Therefore, they have insisted that the CD should deal with the issue in this context by bringing into being another ad hoc committee in which parallel negotiations on nuclear disarmament could be conducted since this was the "highest priority" for the CD.[8] India, Pakistan and Egypt have been among those seeking this as a prerequisite for any progress on the FMCT.

Meanwhile, some others like France and the UK added another twist to the whole issue by demanding that the establishment of a conventional disarmament committee must accompany the establishment of the nuclear disarmament committee. Whatever may have been the motive of such insistence, it made further progress impossible until towards the end of 1999, the G-21 countries relaxed the linkage between the FMCT and simultaneous nuclear disarmament negotiations in an ad hoc committee. They settled instead for informal discussions and exchange of views on nuclear disarmament in ad hoc working groups. Progress, however, remained elusive because of lack of agreement over the precise language of the mandate of a working group!

The NMD Angle

The other issue that has impacted upon the negotiations of the FMCT, or the lack of these in the CD relates to the US NMD plans. The likelihood of NMD deployment in violation of the ABM Treaty negotiated with the USSR in 1972 has brought to the fore a new set of issues and concerns. The US has justified the need for putting into place a missile shield, citing the threats arising from "states of concern" like North Korea, Iran or Libya. However, the Russians, as also the Chinese, read more into the US NMD effort. They perceive it as a means to lower the efficacy of their own nuclear deterrent. Consequently, they have criticised the proposed US endeavour as an unacceptable effort designed to achieve strategic superiority in the 21st century at the cost of disrupting global and regional strategic balance and stability, besides triggering off a new round of the arms race.

A reinforcement of its nuclear status became even more important to Russia as the US resolve to go ahead with the deployment of the NMD gained momentum after the entry of President Bush into the White House. In Russian thinking, its nuclear weapons are of crucial importance for the country for two main reasons. Firstly and more obviously, to fill an internal void and give Moscow the requisite strength to pursue domestic policies of the kind that it deems fit. The nuclear weapons have been perceived as a sort of a shield to insulate external criticism at a time when the other determinants of national status and international prestige are found wanting. This was amply evident in the manner in which references to Russia's nuclear strength were made while responding to international criticism of the Russian policy in Chechnya.[9]

At the same time, the Russian nuclear armoury also serves an external need. It fortifies an otherwise weak nation against the relative strength of some of the other international players that Moscow perceives as security concerns. For example, Russia does feel threatened by the increasing tendency on the part of the US and its allies to orient international relations towards unilateralist actions and a consequent undermining of the UN. Placed in such a matrix of an insecure domestic situation and a complicated and seemingly threatening international environment, Russia has perceived immense political and military possibilities in its nuclear arsenal.

By the turn of the millennium, as NATO remained as threatening as ever to Russian security with its eastward expansion and its reiteration of the first use of nuclear weapons in its new security concept put forth in 1999, Russia too reaffirmed its nuclear arsenal as a "key element" of its security. It "vowed to continue to maintain and develop, including through sub-critical testing, the strategic arsenal."[10] The Russian president also signed a decree outlining a concept for the development and deployment of tactical nuclear weapons, ostensibly in a bid to overcome the deterioration in Russian conventional weaponry.[11]

While it may be true that the present economic state of Russia does not allow it to maintain a large nuclear arsenal, if financial conditions permit, Russia has mentioned the possibility of redeploying tactical nuclear weapons at its borders, and having more MIRVed nuclear weapons in order to render US national defences ineffective. It may be recalled that the Duma ratification for START II came about not only after a long delay but Russia also made its implementation contingent on US compliance with the ABM Treaty. Moscow had then mentioned the possibility of its breaking free from its commitments under START II, especially with regard to the limits it imposes on MIRVs, if the US was to go ahead with NMD deployment in violation of the ABM Treaty. Since then, the US has withdrawn from the ABM Treaty and there is an ostensible sense of reconciliation on the part of the Russians over the move. While expressing regret at the US decision, Russia, for the time being has responded positively to US overtures and advances in smoothening their bilateral relations. NATO and Russia too have drawn closer, more than ever at any time in history, and there is a joint concern and resolve to fight against terrorism, the most potent threat of these times. This closeness, however, does not in any way reduce Russia's reliance on its own nuclear arsenal; neither can it stop it from continuing to develop its own counter-measures for the US NMD. In the process, however, the FMCT remains removed from the US-Russia radar screen for the moment.

China, too, has been particularly vocal about its dissatisfaction with NMD deployment and the consequent Russian moves to strengthen its own nuclear deterrent. The Chinese apprehensions on both these counts have been amply articulated in several international fora. For instance, at the NPT RevCon in 2000, the Chinese representative, Ambassador Sha Zukang, lamented the continuance of the Cold War mentality. He said, "Military alliances have been strengthened, hegemonism and power politics manifest themselves from time to time, and a small group of countries advocates the so-called 'humanitarian intervention', by which they violate state sovereignty."[12] He further went on to elaborate upon the negative consequences of these developments as visible in "the increase of military factors in international relations and a decline in the sense of security among countries...(that) seriously undermine the trust and cooperation in the field of international arms control."

China has been quick to point out that the US NMD deployment would lead to the eventual weaponisation of outer space. Apprehensive of what this would signify for its own nuclear deterrent, China has not missed any opportunity to criticise the US initiative. In fact, in the FMCT negotiations, it seems to have discovered an effective tool since it realises the American keenness to have China as a party to the treaty, especially if it wants to rope in India and Pakistan. Consequently, Beijing has seen merit in linking its support for the FMCT to the establishment of an ad hoc committee to negotiate the prevention of an arms race in outer space (PAROS).[13] For the past four years, progress in the CD has remained stalled on this issue because of the Chinese unwillingness to move unless there is simultaneous movement on the PAROS. It has made the negotiations on the fissile ban conditional upon an agreement on a programme of work for the CD that is balanced and comprehensive, i.e. one that would give equal weightage to the issues of militarisation of space and nuclear disarmament. Rather, it considers the prevention of the weaponisation of outer space to be a task even more urgent than the FMCT negotiations. At the NPT RevCon held in May 2000, it did not even agree to issue a call for a collective moratorium on the production and stockpiling of fissile material, even though it is widely believed to have stopped production of its fissile material some years ago.[14] Instead, in a clever move, it linked the call for a moratorium on fissile material production with one seeking a similar ban on deployment of weapons in outer space.

The US, on the other hand, is the sole country blocking outer space negotiations since it feels that the time is "not ripe" for negotiations on either the outer space or disarmament issue.[15] Rather, the US has criticised this linkage as "simply a poorly disguised effort to block FMCT negotiations altogether."[16] Consequently, all through 1999 and 2000, the CD continued to remain deadlocked over the competing negotiating priorities of the US and China.

On the whole then, arms control, including the FMCT, does not appear to figure in the scheme of things as being planned out by the major nuclear powers. In fact, the differences in their priorities and perceptions are best reflected in what has been the fate of the FMCT negotiations until now. Whether the treaty is ever concluded and the form that it takes will also ultimately depend on how the major international players look upon arms control in the coming years.

India and the Proposed FMCT

In principle, India has expressed its support for the early conclusion of a non-discriminatory treaty that would tackle the issue of fissile material in its totality. However, there are two points that it emphasises with regard to the treaty. First, that the FMCT be a "cut-off" treaty restricted only to future production. In fact, India, at one stage, had made it explicit that it would not participate in the fissban negotiations unless the treaty was restricted only to future production. At the start of the CD's 1997 session in Geneva on January 21, 1997, India let it be known that the country would endorse the FMCT only if the CD stuck to the mandate as given by the UNGA in 1993. According to New Delhi, this mandate upheld that "the FMCT must not be seen in the retrospective sense. It has to be viewed as a prospective measure which prohibits the future production of unsafeguarded material."[17] A few months later, while addressing senior scientists of the Bhabha Atomic Research Centre in Mumbai on May 31, 1997, then Prime Minister I.K. Gujral had asserted that India would not sign the FMCT unless its precondition was met.

The second qualification of sorts that it has attached to the FMCT is that the treaty be placed in the wider context of nuclear disarmament. That it should be seen not as an end in itself but as one step on the road leading to the eventual realisation of an NWFW. Prime Minister Vajpayee said as much in his address before the UNGA in 1998 and in 2000. However, on both occasions, he predicated India's participation in the negotiations "in good faith in order to ensure a treaty that is non-discriminatory and meets India's security imperatives."[18]

Even though little headway has been made on the treaty in the CD, it would be of utmost importance for India to understand the implications of the FMCT as and when it is concluded. India is a country with an ambitious nuclear power programme and at the same time is also in the process of building a minimum credible deterrent. The FMCT would carry certain implications on both these counts. It must be understood that unlike the CTBT that basically involves curtailing only weapons related activities, the FMCT deals with fissile materials and their production technologies that can be subjected to dual use. Therefore, in its effort to address the issue of non-proliferation, the FMCT could affect many related interests impacting on civilian nuclear technologies. The overlap between the bureaucracies and scientific establishments responsible for the design and development of nuclear weapons and those in charge of commercial nuclear power cannot be wished away. In fact, the possibility that nuclear establishments could use nuclear power as a cover behind which to maintain readiness to resume nuclear weapons production was articulated way back in 1946 by J. Robert Oppenheimer, chairman of the General Advisory Committee of the Atomic Energy Commission. While speaking in the context of a convention on international control of nuclear weapons and nuclear disarmament, he had said, "We know very well what we would do if we signed such a convention: we would not make atomic weapons, at least not to start with, but we would build enormous plants, and we would call them power plants...we would design these plants in such a way that they could be converted with the maximum ease and the minimum time delay to the production of atomic weapons..."[19]

This relationship between the two technologies is only likely to further increase by proposals that encourage the production of tritium for weapons in power reactors, or projects that seek conversion of surplus weapons plutonium into reactor fuel. The existence of this duality makes it important that India must be very clear on what the treaty implies in its definition, its exact scope, and its verification provisions.

Definitions

According to the most commonly accepted definition, the term fissile material includes those materials that are fissionable when irradiated by both slow or thermal neutrons.[20] The IAEA Statute defines special fissionable materials as plutonium-239, uranium-233, uranium enriched in the isotopes 235 or 233 which means uranium containing either or both of these isotopes in an amount such that the abundance ratio of the sum of these isotopes to the isotope 238 is greater than the ratio of the isotope 235 to the isotope 238 occurring in nature.[21]

One crucial question that has been raised in this regard is whether the definition of fissile materials should be restricted to only "weapon grade" materials such as enriched uranium and plutonium, or whether it should also include "weapon usable" materials such as tritium as well. Technically, tritium is not a fissile material, but is used to enhance the yield of nuclear weapons. Logically speaking, if the FMCT is truly to be a step toward capping further arms production, then it must include tritium as well. But the US has sought to keep this material off the bounds of the treaty. It has upheld the viewpoint that bringing in tritium under the fissban would make the treaty a "weapon ban" and not a fissile material cut-off. In fact, even though it does not produce fissile material any longer, in December 1998, Energy Secretary Bill Richardson designated three Tennessee Valley Authority commercial nuclear reactors as the "preferred facilities" to produce tritium for nuclear weapons. Tritium decays at the rate of about five per cent per year. So, the US is likely to need new tritium by 2005, if weapon stocks remain at START 1 levels. Even if they are reduced to START II levels, the US would still need tritium by 2011.[22]

The question of tritium must also be considered from the viewpoint of the Indian nuclear arsenal build-up. In an article written in January 1998 and published in *Jane's Intelligence Review,* T.S. Gopi Rethinraj had emphasised the "importance of tritium as a strategic material in the creation of thermonuclear weaponry". Having carried out a thermonuclear explosion in May 1998, and with a demonstrated capability of producing fusion bombs, India would need tritium to boost the yield of the weapons. In fact, all modern nuclear weapons use the deuterium-tritium gas fusion boosting of fission devices and India is believed to have used the same in one or more of its low yield tests. If this is true, then, owing to the technical advantages of this technology, it would be possible for India to replace weapon grade plutonium with fuel grade plutonium as fissile material in a weapon. This could have significant implications for the number of nuclear weapons that India can build since India's stockpile of fuel grade plutonium is substantially more than that of weapon grade plutonium. The latter is believed to be enough for only 100 weapons at present. Meanwhile, it is estimated that some 6,000 kg of plutonium has been discharged from India's power reactors of which about 2,000 kg has been separated from spent fuel. This could be enough for making nearly 400 weapons.

While there is presently no indication that tritium would be included in the definition of fissile material under the FMCT, India must clearly understand the implications of its inclusion, or non-inclusion.

The second definition to which proper attention must be paid is "production of fissile material". Questions such as which exact activities should be considered as engaged in the "production of fissile material" under the treaty could have far-reaching implications. This would determine whether only reprocessing and uranium enrichment facilities—the two primary pathways to fissile material production—should be considered as engaged in "production" or whether other facilities that could be potential sources of fissile material, such as nuclear reactor operations and spent fuel storage units, too should be treated as fissile material production facilities.

It is very important to have clarity on these definitional aspects since the scope of verification shall be determined on this basis. While logically speaking, all facilities and activities that could be used to produce or divert weapons usable fissile material should be subject to FMCT verification, ambiguities arise due to the dual use nature of such activities. For instance, natural uranium fuelled reactors being used for electricity generation could also become plutonium production facilities. Similarly, tritium production reactors can be used to produce plutonium or even highly enriched uranium. But unless tritium is included in the treaty, these facilities would be outside verification purview.

Scope

As has already been mentioned earlier in the chapter, the debate over scope has basically been polarised over whether the fissban should apply only to future production or whether it should also tackle the issue of existing stockpiles. Prominent questions pertaining to scope that are yet to be clearly answered are: should stocks of existing fissile material be declared under the treaty? Should stocks declared excess and subject to international verification be included in the treaty? Should preexisting facilities that contain fissile material be declared under the treaty? Should the clean-up or recycling facilities for existing fissile material be included in the treaty? And lastly, how should commercial reprocessing of plutonium be viewed?

India has not favoured the inclusion of existing stockpiles within the treaty ambit, since naturally it would impinge upon the size of the nuclear deterrent that India might want to, or be required to, build in the future. However, Pakistan has insisted in the negotiations that a cut-off of future production should be accompanied by declarations of existing fissile material stockpiles and a schedule for their transfer to safeguards. While Pakistan has premised this demand on the ground that it would lead to the elimination of asymmetry in the possession of fissile material stockpiles by various states, India cannot afford to miss its implications. For one, it would actually lead to the removal of asymmetry, and for another, Pakistan could also be using it as a delaying tactic to gain time to build its own stockpiles while also finding a way out of its purely HEU based weapons programme.

This leads one to the question of plutonium. For India, the fine text on plutonium in the treaty must be carefully read since the material has been designated to play a crucial role in the power programme as envisioned by Bhabha and especially as regards its role in the thorium cycle. It is for this reason that in the negotiations, India has sided with the UK and France to let the FMCT allow plutonium production for civilian purposes. But this can have its own dangers given the mercurial nature of this material. It has been established that once plutonium has been separated from the highly radioactive spent fuel, it would not be difficult or time consuming to convert it into metal, the chemical form most appropriate for weapon use. This could even be done in facilities that are simple and small and, hence, could easily escape detection. Pakistan's attempts at developing its plutonium stockpiles must then be considered from the angle of the security threat they could impose.

Commercial reprocessing of plutonium is another area on which opinion has been divided. Countries like Russia, Japan and even Germany, Switzerland and Belgium that have a substantial civilian nuclear industry have expressed interest in starting an extended plutonium economy, including fast breeder reactors.[23] Russia rejects the idea of disposition of excess weapons plutonium without extracting a commercial benefit from it. The US, on the other hand, after having renounced civilian plutonium recycling since the days of the Carter Administration, is more concerned about the implications of such activity.[24] India too would need to consider this issue in order to articulate an informed response based on its security and commercial requirements.

Of course, another very important issue as regards the scope of the treaty pertains to whether the production of fissile material as a by-product of electricity production or other civil purposes in nuclear reactors that are not under international safeguards should be exempted from safeguards in states where all nuclear material is not subject to IAEA safeguards. This is of particular relevance to India since not all Indian nuclear facilities are open to IAEA supervision. In fact, while for the five NWS one primary objective of the FMCT is to bring such facilities in NPT holdouts under IAEA safeguards, India will need to consider this question carefully in order to weigh its security implications.

Verification

Effective and reliable verification would constitute the heart of the FMCT. In fact, the acceptance of the treaty would hinge on whether countries see its verification provisions as sound enough to detect and stall cheating, while at the same time not being unnecessarily intrusive or abrasive to state sovereignty. Simply put, the FMCT would have to cover three classes of facilities. These would include declared shutdown production facilities, declared operating production facilities and a look out for any undeclared production facilities.

During the limited negotiations held so far, many options have been proposed ranging from focussed to wide verification. The former envisages concentrating on only sensitive fissile material production facilities such as enrichment and reprocessing facilities, on fissile material produced after the treaty's entry into force and on facilities where these materials are present. A wide scope verification mechanism, meanwhile, proposes the inclusion of even less sensitive civil facilities such as fuel fabrication plants and civilian power reactors.

Verification would have to be realised through safeguards similar to those administered by the IAEA. At this juncture, it would also become necessary to address the issue of India's nuclear status. In case the same verification provisions are applicable to all countries, there would be little problem. But if their nature is different for the declared NWS, then the *de facto* nuclear, but *de jure* non-nuclear status of India and Pakistan could pose a difficult proposition.

The applicability of safeguards to the NWS could pose further problems since sensitive production facilities on which these safeguards would apply could be co-located with other sensitive facilities where military activities such as the assembly of nuclear weapons could be in progress or where undeclared stockpiles from past production are being held. Inspections could then cause worries about leakage of sensitive information. The verification system would also have to be adequately supported by the establishment of a material accounting system by the state that is subject to on-site inspections in order to determine the accuracy and veracity of the accounting system. Other verification measures that could be used would be challenge inspections, environmental monitoring and remote sensing involving satellite imagery.

Conclusion

From the issues raised and discussed in the above sections, it is amply clear that the negotiations for a fissile material cut-off are going to be a complicated and acrimonious affair. Divergent viewpoints arising from country apprehensions on various accounts do not augur well for the talks. Meanwhile, the issue of when the negotiations will actually meaningfully begin remains embroiled in the larger perspectives on arms control. Contemporary international relations do not seem to accord much priority to arms control endeavours. Rather, they have been reduced to being used as leverages in inter-state relations.

Such a scenario does not bode well for international security. Nor may it be taken as a happy prospect for India. While the Indian scientific establishment must make use of this lull in nuclear diplomacy to closely contemplate the issues raised in the context of the FMCT, the bottom line remains that an overall reduction in international tensions cannot but be in India's long-term interests. Therefore, in order to promote international security, India must continue to focus on the need for effective and irreversible arms control as a short-term measure, and universal nuclear disarmament as a long-term objective. Within this context, support for the FMCT must continue, with special efforts during the stage of negotiations to raise pertinent issues and shape the treaty into one that suits India's national interests best.

Notes

1. United Nations General Assembly, 48/75L, Consensus Resolution of December 16, 1993.
2. CD/1299, March 24, 1995.

3. Brian G Chow, Richard H. Speier, Gregory S. Jones, *The Proposed Fissile Material Production Cut-off: Next Steps,* RAND Report (Santa Monica: RAND, 1995), p. 6.
4. The divergent opinions are naturally also reflected in how countries refer to the treaty. While India continues to use the acronym FMCT since it argues that the treaty must only prohibit future production, other countries have used different terminologies as per their country positions. Interestingly, Pakistan, that argues for the inclusion of existing stockpiles, prefers to use "FMT" and leaves out the "cut-off". Several others also use the more general term "fissban" so as not to prejudge the issue of scope and stocks.
5. CD/PV/681, June 9, 1994.
6. Statement of Munir Akram, ambassador of Pakistan to the CD, August 11, 1998.
7. Report of Ambassador Shannon to the CD, March 23, 1995.
8. G-21 Statement at the CD, August 11, 1998 http://basicint.org
9. "Russia Calls NATO 'Cynical' Over Chechenya", *Reuters* (Moscow), December 3, 1999.
10. "Russia Stresses Nuclear Weapons as START II Languishes," *Disarmament Diplomacy,* issue no. 37, August 1999. http//www.acronym.org
11. "Yeltsin Signs Decree on Tactical Nuclear Weapons," *Arms Control Today,* April/May 1999.
12. Statement of the Head of Delegation of the Government of the Republic of China at the Review Conference, New York, April 24-May 19, 2000, as reproduced in *Strategic Digest,* June 2000.
13. The fact that the CD works on the principle of consensus gives each country a veto power over all decisions and this has been termed as the "power of linkage."
14. Rebecca Johnson, "NPT Notes: News From Behind the Scenes", http://www.basicint.org
15. Statement made by US Ambassador Robert Gray at the CD on February 17, 2000, and as quoted by Wade Boese, "CD Deadlock Continues as US and China Square Off," *Arms Control Today,* March 2000.
16. Jenni Rissanen, "Silence and Stagnation as the CD Concludes Fruitless Year", *Disarmament Diplomacy,* issue no. 50, September 2000. http://www.acronym.org
17. *The Times of India,* February 17, 1997.
18. PM's speeches at the UNGA, reprinted in *Strategic Digest,* November 1998, and September 2000.
19. Robert Oppenheimer, "International Control of Atomic Energy," in Morton Grodzins and Eugene Rabinowitch, eds., *The Atomic Age: Scientists in National*

and World Affairs (New York: Basic Books, 1963), p. 55, as quoted by Arjun Makhijani in "Achieving Enduring Nuclear Disarmament," *Science for Democratic Action,* vol. 6, no. 4, September 1999, IEER publication.

20. Brian G Chow, Kenneth A. Solomon, "Limiting the Spread of Weapon-Usable Fissile Materials," RAND Report (Santa Monica: RAND, 1993), p. 1.

21. IAEA Statute as reproduced in David Fischer, *Towards 1995: The Prospects for Ending the Proliferation of Nuclear Weapons* (Aldershot: Dartmouth Publishing Company, 1993).

22. Charles D. Ferguson, "TVA Gets the Add," *Bulletin of Atomic Scientists,* March/April 1999.

23. For more on this issue, see http://www.rz.uni-frankfurt.de/~schaper/seminar/nucmat.htm

24. The US has followed a double approach for the disposal of its own excess weapons plutonium. It has resorted to direct disposal of high-level waste through vitrification, and, secondly, also burned it up as mixed oxide (MOX) fuel in light water reactors.

6

Ballistic Missile Defences: Whither Nuclear Deterrence?

Manpreet Sethi

If nuclear deterrence was the primary philosophy underlying international strategic stability during the Cold War years, then ballistic missile defence (BMD)[1] has emerged in the contemporary nuclear lexicon as the new buzzword. The US determination to operationalise a national missile defence (NMD) system to protect itself from errant incoming missiles launched from a state of "proliferation concern",[2] as also a theatre missile defence (TMD) system for its forward-deployed troops, bases and allies in Europe and Asia, threatens to throw the traditional concept of nuclear deterrence out of gear. Coupled with the now overtly outlined strategy of preemption, the US moves in this direction shall impinge upon the security concerns of every nation and, hence, compel each country to assess its potential ramifications for its own security.

Even though the NMD had emerged while Bill Clinton was president, it had not then acquired as much invincibility as it did once the Republicans took office. In fact, during his last months in the White House, Clinton deferred the decision on the NMD and showed himself to be more amenable to the opposition being expressed to NMD deployment from other countries. President Bush and his team of security advisers, however, had endorsed the concept right through their election campaign and it was soon after his ascendancy to power that he let it be known that the US would be going ahead with its missile defence programme irrespective of international public opinion on the subject. In a landmark speech delivered at the National Defence University (NDU) on May 1, 2001, President Bush interpreted the US pursuit of the NMD as a positive development since, "Deterrence can no longer be based solely on the threat of nuclear retaliation."[3]

Since that landmark pronouncement, US officialdom has consistently endorsed the inevitability of the need for missile defence in the wake of the uncertainties inherent in the present day international scenario. On more than one occasion, Lt Gen Ronald Kamish, the director of the US Defence Department's Ballistic Missile Defence Organisation (BMDO), has emphasised the inability of superior US conventional and nuclear weapons to deter countries that are bent on harming/destroying US citizens and assets. Citing the case of the Cuban missile crisis, he has underlined the fact that though the US nuclear weapons were able to deter the Soviets, they did not have the same effect on Fidel Castro who was "'willing to sacrifice his country and 'die beautifully' in the process."[4] US-identified rogue states and leaders in the present times are believed to harbour the same mentality and are perceived to be undeterred by the offensive military might. This sums up the rationale for a defence shield to neutralise the reality of missile proliferation and the risk of ballistic missiles.

The bolt-out-of-the-blue terrorist attacks on the twin towers of New York City and the Pentagon only seem to have strengthened the US' paranoia and resolve for the erection of an NMD. While it may be argued that a missile defence would do little to safeguard national security against non-state suicidal actors, the US military-industrial-congressional complex has managed to imbue the NMD with a momentum of its own that appears impossible to stall through logic. Rather, it is complemented by a persecution complex that the US has long suffered from owing to its lifestyle and levels of technological and material advancement. Even as far back as in 1948, George Kennan, one of the key architects of post-World War II US policy had stated:

> ...we have about 50% of the world's wealth but only 6.3% of the population. This disparity is particularly great between ourselves (sic) and the peoples of Asia. In this situation, we cannot fail to be the object of envy and resentment. Our real task in the coming period is to devise a pattern of relationships which will permit us to maintain this position of disparity without positive detriment to our national security. To do so, we will have to dispense with all sentimentality and day-dreaming; and our attention will have to be concentrated everywhere on our immediate national interests.[5]

As one can clearly sense, even half a century later, nothing has changed. If anything, the disparities between the US and other societies have only increased, not the least because of US designs. Driven as always to uphold its national interests, and equipped to do so by an unparalleled military might and technological edge, the US has now identified adding a missile defence to its already existing formidable offensive capabilities as another means of safeguarding its national security.

However, not everyone, not even within the US, is convinced of the usefulness of this exercise. Most countries, and not just Russia and China, but also US allies, are sceptical about the efficacy of the BMD and apprehensive that its repercussions on international security and stability may not be all benign.[6] The chain reaction that may be set into motion as others either stumble down the same path or search for other counter-measures is only likely to further complicate the grid of international relations.

Issues Confronting India

Given the profound implications as are disclosed in detail in the following sections, it is of crucial significance that India should accord adequate attention and thought to this issue. President Bush's speech at the NDU that provided a sketchy description of the main planks of the new security framework was, in a rather uncharacteristically quick response, welcomed by the Indian officialdom. However, the initial enthusiasm had waned within a few days as official Indian statements became tempered with greater caution. In fact, by the time of the visit of US Deputy Secretary of State Richard Armitage to New Delhi in the latter part of May 2001, with the stated objective of explaining the broad contours of the US initiative, India had shifted to a more cautious approach. Since then, and as was evident during External Affairs Minister Jaswant Singh's visit to Russia in early June 2001, India appeared to have further watered down its support for the US proposed global security set-up based on a renunciation of the ABM Treaty and the construction of the NMD.[7]

However, as US resolve has strengthened in this regard and especially after it served notice to withdraw from the ABM Treaty in December 2001, India, as also the rest of the nuclear countries, have been swamped by a sense of resignation. From actively opposing the US move in the initial years, the NWS seem to have reconciled themselves to the inevitability of this development and the actual US withdrawal from the ABM Treaty at the end of the six-month notice period evoked nothing more than rather staid and muted responses of regret.

Given Indian aspirations to play a significant role in the international arena, and considering the potential of the NMD to reshape concepts of global security and arms control, it would be prudent that the country should arrive at a considered position on this issue. In order to do so, India will have to carefully study how the NMD would impact on its own security by carefully examining all the attendant risks that could be generated in the likely responses of China, and by extension, of Pakistan, the two main security concerns of India.

A methodical analysis of this kind would also better equip India to convey its concerns to the US. Washington has included India in its list of allies and friends with whom it now engages in "real consultations" to discuss the "common responsibility to create a new framework for security and stability that reflects the world of today."[8] It is now up to India to articulate a coherent and well thought out response to the US plan and make use of the "consultations" to ensure that the new framework that may emerge takes adequate cognisance of India's security concerns.

This would nevertheless require addressing questions such as: what exactly does the BMD entail? How are India's adversaries likely to react to it? Should India be satisfied that the NMD will be accompanied by a reduction in US nuclear forces? Will these reductions necessarily lead to similar cutbacks in the arsenals of others? Would not China respond by increasing the numbers and sophistication of its nuclear arsenal and would this not have an effect on the Indian credible minimum deterrent? Should India extend support to the US initiative even at the risk of an increase in its own threat perceptions? Can India expect the US to come to its rescue in times of need? Should New Delhi endorse the plan for the sake of forging a closer relationship with the US, especially since Indo-US relations are presently on an upswing?

Finding answers to these queries would mean passing through a number of as yet dark alleys. The relative "newness" of the idea of juxtaposing a defensive framework with a reduction in nuclear arsenals, and its formalisation at a time when international relations are being transformed due to globalisation, revolution in military affairs and information technology does not make the task of assessing the possible ramifications of the BMD any easier. This chapter attempts to explore some of the major issues that the BMD shall throw up for India. It serves to do so by first providing a brief introduction to the present strategic context in which missile defences are being projected by the US as the most viable means of meeting the emerging threats to international security. It then examines the likely effects it would have on the regional and international nuclear landscape from a primarily Indian perspective. The conclusion sums up how India could best respond to these developments in order to maximise gains and minimise costs.

The Context

On May 1, 2001, US President, George Bush, Jr. belaboured to drive home a point. "Today, the sun comes up on a vastly different world. The wall is gone and so is the Soviet Union...Yet, this is still a dangerous world, a less certain, a less predictable one."[9] Indeed, with the disappearance of the Iron Curtain, the US worldview changed dramatically. For nearly forty long years, Communist Soviet Union, with its formidable conventional and nuclear might had remained the single most crucial defining feature for the United States. All its policies—security, military, political and economic—had hinged on two objectives: one, to undercut Soviet influence where it existed; and two, to prevent it from spreading to newer regions. In the face of the Soviet threat, all others paled into insignificance.

However, the disappearance of the USSR, and an improvement in strategic relations between the two superpowers, allowed the US to discover and concentrate on new security threats. The most palpable and real among these was identified in the proliferation of WMD, including missiles capable of threatening the American mainland. A bipartisan commission instituted in the second half of the 1990s reported that "the threat to the US posed by these emerging (missile) technologies is broader, more mature and evolving more rapidly than has been reported in estimates and reports by the Intelligence Community."[10] The commission, known more commonly as the Rumsfeld Commission, after its chair, Donald Rumsfeld, who is now the secretary of defence in the Bush Administration, identified countries such as Iran, Iraq, North Korea and Libya as states of proliferation concern. Its findings that were put out in July 1998 appeared to have been vindicated when within weeks of the report, North Korea, in a surprise move, test-fired a long-range version of its Taepo Dong 1 missile on August 31, 1998.

While it may take several years before Pyongyang,[11] or Tehran or Baghdad, would be able to deploy operational missiles capable of striking the US mainland, the spread of WMD technologies to these countries which are not particularly friendly to the US and which Washington describes as the "world's least responsible states"[12] is being perceived as the biggest threat to the country and its allies. This then, has prompted the US to reorient its national security strategy from one that was geared to meet the threat from a huge Soviet arsenal to one that seeks to provide protection against states that might have a "small number of missiles," but "for whom terror and blackmail are a way of life."[13]

Of course, every country has the right to undertake periodic threat assessments and arrive at means that it considers the most appropriate to defend itself.[14] During the years of the Cold War, when the two superpowers were nuclear armed to the teeth, offence was deemed as the best form of defence. Mutual vulnerability was considered crucial to hold nuclear deterrence in place since the near-certainty of assured destruction was sure to keep the use of nuclear weapons in check. However, with the change in the nature of threats, the US has perceived a need to change the nature of its security strategy from purely offence to a mix of offence and defence. During his term, President Clinton signed this into law as the National Missile Defence Act. It confirmed the US policy to "deploy, as soon as it is technologically possible, an effective National Missile Defence system capable of defending the territory of the United States against limited ballistic missile attack."[15] Since moving into the White House, President Bush has lent complete support to the NMD and in his speech rendered at the NDU in May 2001, declared it the cornerstone of future US security strategy.

The emphasis now is on constructing a new security paradigm. As President Bush put it, "We need new concepts of deterrence that rely on *both offensive and defensive forces*. Deterrence can no longer be based solely on the threat of nuclear retaliation."[16] He has hinted at three main planks of the new security architecture: active non-proliferation; counter-proliferation; and defences. The amount of importance that is accorded to each of these and how they are pursued is, however, still not clear. As far as non-proliferation is concerned, it has been a long held objective of the US. It has been pursued through the establishment of a regime erected on pillars such as the NPT, the CTBT and the FMCT. Nevertheless, the success record of the regime is not blemish free. Rather, it is perhaps a failure of the NPT that nuclear weapons and/or their technologies have today become available with countries that are NNWS parties to the treaty, and, hence, legitimately prohibited from acquiring or developing the same. In fact, the acquisition or development of WMD technologies has been largely a result of collusion of the NWS themselves. Selective proliferation has been blatantly pursued by some of the NWS, leading to the contemporary situation in which the US now perceives a threat from the small WMD capabilities of states, all of which, ironically, are NNWS of the NPT. This raises the fundamental question of whether the NPT, in spite of its permanent extension, will prevent future proliferation considering that 188 countries are members of the treaty and only three (all of them nuclear anyway) are outside it? As regards the CTBT, the denial of ratification to it by the US Senate has virtually spelt its death-knell.[17] The FMCT is nowhere near conclusion. In fact, negotiations remain stalled mainly owing to the different perceptions of how the NMD would eventually lead to the next phase of militarisation of outer space.[18] In this context, what direction future US non-proliferation initiatives will take remains unclear.

Neither is the strategy of counter-proliferation all in place. While the concept has been doing the rounds of the US strategic circles from the early 1990s, counter-proliferation is still fluid and vaguely defined. It basically encompasses a fight against proliferation through the launch of preemptive strikes aimed at a surgical removal of WMD facilities of a country perceived to be an enemy of the US. The initiative assumes that offensive military operations can neutralise or roll back the threat of strategic weapons proliferation. However, the two main operations that sought to apply the strategy of counter-proliferation in the late 1990s are not considered clear successes. In 1998, US cruise missiles were fired against a suspected Sudanese chemical weapons plant that later was discovered to be a pharmaceutical factory. Similarly, US attacks against Iraqi chemical, biological and missile production plants until 2002 had still not enabled the US to confidently assert that the threat from Iraq has been completely neutralised.[19] On the contrary, at the time of writing this chapter, the US is already fighting in the UN to seek sanction for a military operation against Iraq.

Given the past record of the limited efficacy of the strategy of counter-proliferation and the possibility of its implementation on false assumptions, it arouses scepticism, and even apprehension, amongst other countries. How far wide and in what manner the ripples of this spread across the world, is yet to be seen. For the moment though the US is not unduly bothered by the possible repercussions of its moves and is focussed instead on erecting and perfecting the third leg of the new triad, namely missile defence.

Ballistic Missile Defence

Predominant wisdom on missile defence during the Cold War years was based on two broad assumptions. One, that a national missile shield that could blunt a large-scale attack had a destabilising influence on the strategic balance. It increased the risk of war and eroded one's faith in one's own nuclear deterrent. In other words, it carried the risk of one country triggering off a war because it felt secure under its own missile cover and not afraid of retaliation.[20] Second, it was presumed that the deployment of such defence systems would trigger an arms race in the upgradation of offensive capabilities in a bid to attack the other's perceived sense of invulnerability.[21] Besides, these two assumptions, of course, on another front, the technological feasibility of the BMD was also in doubt, especially since the threat to be catered for was not a country with a few missiles but the USSR that had thousands of warheads and a formidable delivery capability. Then, in view of the technical challenge and the immense costs it would have involved, Washington preferred deterrence to defence.

It was in the context of these parameters that the US and the erstwhile USSR had signed the ABM Treaty in 1972. It prohibited the deployment of strategically significant ballistic missile defences that could erode the other's nuclear deterrent. Both sides agreed to remain mutually vulnerable to each other's nuclear weaponry and hoped that this would prevent a nuclear exchange. Of course, Article III of the treaty permitted each side to deploy an ABM system at two sites (later amended to one site in 1974), each limited to 100 interceptors. During the years of the Cold War that were marked by a lack of trust on either side about the other's intentions, this treaty did serve its purpose.

However, in the present context, three main developments are cited by US strategic experts as providing reason enough for a rethink on the ABM Treaty. The first one is the end of the period of hostility between the US and USSR and the gradual blossoming of a cooperative relationship. This development has, in turn, timed well with the emergence of WMD threats from smaller arsenals in rogue states. As the Rumsfeld Commission concluded, "A new strategic environment now gives emerging ballistic missile powers the capacity, through a combination of domestic development and foreign assistance, to acquire the means to strike (the US) within about five years of a decision to acquire such a capability (ten years in the case of Iraq)."[22] The prospective missile arsenals of these countries can only be modest, in low two-digit numbers. These can then be neutralised by an NMD, especially since there have been advancements in missile defence technologies that now make it possible to effectively blunt these threats better than in the past. Naturally then, the US Administration sees sense in deploying this latest technology to address its threat perceptions. It can be expected to safeguard the US' freedom of action even in the face of coercive missile threats from smaller regional powers, while also act as a protective shield in case deterrence fails.

Consequently, in the new post-Cold War context, the concept of NMD has reemerged in a transformed strategic environment, with a new mission, against a new kind of a threat, and equipped with newer advancements in technology. It is also much less ambitious compared to what President Reagan had envisaged in the Strategic Defence Initiative (SDI) in the 1980s. Recognising the limitations of technology, the current NMD system seeks to protect the US against only a small-scale missile attack of no more than a few warheads.

Even this, however, may not be easy and absolutely guaranteed. While the Pentagon has demonstrated the basic feasibility of the hit-to-kill concept, not all tests have succeeded in intercepting incoming missiles. Several technical problems still underlie an effective NMD system[23] that would be required to detect an offensive missile launch, track the missile in flight and finally intercept and destroy the missile or the warhead(s). The director of the US BMDO has stated that the technologies presently under development will only be able to defend against incoming missiles in their terminal stages.[24] However, the US hopes that within the first quarter of the new century, the country would have mastered the capability to build sophisticated ground-based radars, as well as a new space-based missile tracking system that would be able to kill an enemy missile mid-course while it is still cruising towards the target. In fact, the BMDO director dismisses the despondency over missile interception failures by pointing out that other ambitious US programmes such as the development of the Atlas ICBM, the Minuteman-I ICBM, etc too had experienced 10-12 failures in their 2-3 years of flight-testing history. Dubbing these as "successful failures," he exudes confidence on the eventual success of the NMD. To quote him, "Seven hundred years ago, the current wisdom held that sailors who ventured too far from land would fall off the edge of the Earth. Columbus proved them wrong. Seventy years ago, naysayers were saying that rockets wouldn't work in space because there was no air to push against. Robert Goddard proved them wrong. Seven years ago, critics were still saying we couldn't hit a missile in space. We've done it."[25]

Indeed, pronouncements such as these demonstrate the determination of the present US Administration to press ahead with the NMD. Not technical failures, and definitely not public opinion, domestic or international is likely to hold them back. With an annual budget of $4 billion and 500 employees working to turn rocket science technology into functioning systems, the US is poised for NMD deployment. In fact, the first systems to provide close-in defence to intercept a short-range missile in its terminal phase that include the army's PAC-3 (Patriot Advanced Capability 3) are near operational. Within a few years, it is expected that the Navy Area system that builds on existing fleet air defence capability would be ready for deployment. Both these systems that are limited to defending a relatively small area, would be eventually complemented within this decade with the army's THAAD (Theatre High Altitude Area Defence system) and the navy's Theatre Wide systems.

BMD Implications for India

Even as the US remains engaged in the research and development of a sophisticated NMD, other countries will have to be concerned about what these advances would mean for nuclear deterrence in general, and their national security in particular. From the Indian perspective, it is essential to look at two issues:

(a) Does India share the US threat perceptions?

(b) Does India consider the US' *modus operandi* as the best way of dealing with these threats?

Significantly, countries that have been enumerated as states of proliferation concern by the US are not essentially perceived as posing any direct threat to India. India's security concerns primarily emanate from its own neighbourhood. In fact, India happens to live in a region where WMD proliferation has been an acute and worrisome reality for several decades now. China has been in possession of a fairly sophisticated WMD arsenal for the past several decades. For India, this has constantly been a source of insecurity, especially given the fact that China lays claims to several hundreds of square kilometres of Indian territory, and border disputes persist. Moreover, India has always been conscious of the growing imbalance of power in Asia due to the rise of an increasingly politically vocal, economically powerful and militarily modernising China and a relative decline in the economic and military strength of Russia and Japan.

Until 1998, when India finally decided to exercise its nuclear option in favour of weaponisation, the India-China relationship had existed as one that had nothing in common with that of mutual assured destruction (MAD) that the US had shared with the USSR. From the second half of the 1960s to the latter half of the 1990s, India had already cohabited with the fear of a nuclear capability, deliverable not only from its eastern but its western side as well. In fact, the threat from China was further compounded by its lack of compunctions about transferring its WMD weaponry and technology to other states, especially Pakistan, for commercial and strategic gains. Given the turbulent and traumatic history of partition and the continuing border disputes, India-Pakistan relations have never been smooth. In an astute move, China has exploited this situation by setting up a thriving nuclear and missile nexus with Pakistan. In fact, it is owing to this collaboration as well as its ability to tap other sources that Pakistan is estimated in some quarters as having a comparatively surer delivery capability for its modest nuclear arsenal.[26]

Sandwiched between two adversaries that work in close strategic collaboration and confronted with real missile threats, if there is any country that needs an NMD, it is India. However, it has never had the luxury of the requisite financial resources, nor the technological capability to pursue dedicated research and development in this field. In fact, to aspire for one through indigenous development would be unrealistic, at least over the next two-three decades, given India's financial and technological constraints.[27] At the same time, to believe that the US shall be ready to put a missile shield over India on account of the improving Indo-US relations would smack of naiveté. For one, all technological hurdles in the path of an effective missile defence system have not yet been overcome. Despite spending tens of billions of dollars, the US still faces serious technical difficulties with its NMD system. The risk of its frustration by counter-measures such as firing a barrage of missiles rather than one or two at a time, concealing the warhead by simultaneously releasing dozens of decoys or even cooling the warhead's nose cone with liquid nitrogen to foil heat sensors, etc. are practical problems that are yet to be surmounted. Even more significantly, any missile defence system that the US may offer to establish over India will have to be technologically even more sophisticated since it shall have to provide protection against short, medium and intermediate range ballistic missiles that are faster and have shorter trajectories. Naturally, this translates into lesser response time, making their interception even more difficult.

Apart from the technical problems, politico-strategic reasons would also stand in the way of an Indian acceptance of a BMD system from the US. To have itself placed under such a system would essentially mean supporting and sustaining unipolarity. Of course, economically it could make sense to avail of the US offer, when and if it ever comes. But surely the US would extract a price for the protective umbrella, perhaps in terms of a blanket Indian endorsement for all its initiatives as the only globocop. Having transformed itself into an invulnerable state through an NMD coupled with a policy of counter-proliferation and preemption, it is not unthinkable that there could be a natural increase in the American tendency to arbitrarily and unilaterally undertake military intervention in the affairs of other states. Iraq and Yugoslavia already represent two instances in which the US chose to act without UN authorisation.

Naturally, this is viewed as a disturbing trend by all the major powers. Indeed, its implications are not lost on Russia that fears a possible US intervention in Chechnya, nor on China that fears US policy on Taiwan. The trend can be no less worrisome for India, especially in the context of Kashmir. US officials have often referred to Kashmir as a "dangerous nuclear flashpoint". While at present, Indo-US relations are inclined towards mutual respect and friendship, there is nothing to suggest that the relations might not sour at some point in the future, as has happened in the past. It should also be remembered that despite its occasional criticism of Pakistani abetment of militant activities in Kashmir, the US has always stopped short of declaring it a rogue state.

India and the US do not necessarily share congruent threat perceptions. In fact, India must maintain that its security lies in a polycentric world order in which each country has sufficient space for action and manoeuvrability. An issue-based alignment with the US on a case-to-case basis would be a sensible policy to follow instead of providing a blanket endorsement to the US for all its actions in return for a not even absolutely foolproof BMD system. It would also do India well to remember that even long time allies of the US have periodically been troubled by the thought of an eventual US isolationism that could leave them to fend for themselves.

In Europe especially, the NMD has elicited several such expressions of concern. The prospect of a "Fortress America" has, in fact, led some of these countries to buttress their own security systems. A number of European countries are engaged in research and development of TMD systems. There is every likelihood that Europe's nascent TMD systems could be drawn into an overarching global BMD system as is being considered by the Bush Administration. However, it would be naïve to expect in the near future a similar deployment over that part of Asia of which India is a constituent.

Therefore, it might be best for India to continue to be militarily self-reliant, to whatever extent possible. It is already engaged in the process of building a credible minimum nuclear deterrent. Till such time as nuclear weapons disappear from the military and strategic calculations of countries, nuclear deterrence, however reluctantly India may have accepted it, would have to remain the cornerstone of India's nuclear strategy. India's concept of minimum nuclear deterrence would, however, in the present context, be required to accord due importance to two factors: the Chinese response to the NMD; and, the impact of the NMD on non-proliferation and disarmament at the global level.

The Chinese Response

For most strategic analysts in the US, the NMD was initially viewed as being of concern only to the Russians. Given the fact that the NMD deployment required an amendment or abrogation of the ABM Treaty, a bilateral agreement between the US and Russia, the focus initially was mostly on reaching some sort of an accommodation with Moscow on the issue.[28] China received little attention in this context. However, China itself lost no time in factoring the US NMD into its strategic and security calculations.

Extremely conscious of the adverse impact that a reliable missile defence would have on its small strategic deterrent missile force, China vehemently opposes the NMD/TMD.[29] In fact, China's response strategy in this regard encompasses three main elements—enhancement of its nuclear deterrence capabilities through military modernisation; alignment with like-minded countries; and, increase in its bargaining power by holding out threats of proliferation of WMD technologies to countries perceived as states of concern by the US.

China holds that even a limited US missile defence has the potential to undercut its nuclear deterrent. With a limited ICBM force of only 20 odd missiles, China looks upon the US plans of NMD deployment as having the potential to severely constrain its ability to deter the US. Until now, China's strategy of minimum nuclear deterrence has been based on a quantitative uncertainty as regards its nuclear arsenal.[30] But with the NMD, the Americans are likely to be freed of the fear of Chinese retaliation since the shield would be able to cater for the small number of Chinese nuclear weapons that might survive a US first strike. This, from the Chinese perspective, disturbs strategic stability and cannot be a welcome prospect given the disputed status of Taiwan.

Mindful of the 1958 Taiwan crisis, Beijing is averse to any eventuality in which its nuclear vulnerability is exposed and it is left without a retaliatory option. If the US were to become immune to Chinese missiles, it would erode the effectiveness of Chinese nuclear forces and handicap Beijing's use of force against Taiwan, if and ever required, to achieve its reunification. This is perceived by the PLA as a major erosion of its power projection capabilities. Having faced the threat of the use of nuclear weapons in the 1950s, first during the Korean War and then during the Taiwan Strait crisis, the Chinese decision to acquire a demonstrated nuclear capability in 1964 stemmed from a desire to "oppose the US imperialist policy of nuclear blackmail and threats..."[31]

Therefore, an NMD that threatens to curtail Chinese manoeuvrability by virtually disarming it would most likely be met by an increase in the number, sophistication and survivability of missiles in order to saturate US defences. In any case, China's military is engaged in an ambitious programme of nuclear and missile modernisation. Research and development is on to graduate to a new generation of solid fuelled, multiple warhead and road-mobile missiles. The first of these new missiles, the Dong Feng 31, was tested in July 1999 and more efforts are concentrated in this direction to improve the chances of survivability of these missiles. A US National Intelligence Estimate revealed in August 2000 that China could decide to deploy up to 200 ICBMs by 2015.[32] China is also believed to be using the Global Positioning System (GPS) to make significant improvements in the accuracy of its missiles. It has reached an agreement to share the Russian space-based Global Navigation Satellite System.[33] In fact, on the issue of the NMD, Russia and China find themselves on the same side of the fence. At regularly held meetings, both have decried the

US move. Russian arms sales to China have increased and so has bilateral trade. While these trends may be attributed to a totally different set of reasons, they do strengthen Beijing's hand, economically, politically and diplomatically.

In fact, it should be expected that China would use every tool to meet the perceived threat from the NMD. And yet, its actions could prove those wrong who believe that that the actual hidden motive of the NMD may be to pull China into an arms race that would bring upon it economic ruin as happened in the case of the USSR. Two factors negate such a possibility. One, Chinese shrewdness and understanding of international relations; and second, the $100 billion worth trade with the USA. In the light of these two realities, it would be naïve to assume that the Chinese would fall into such a trap. Rather, it should be expected that they would use other means to shore up their defences.

Apart from the steps taken to directly bridge the missile gap with the US, it is also highly probable that Beijing would craft its responses along different lines. For instance, China has hinted at becoming more belligerent and less cooperative on a range of non-proliferation and arms control issues. The Chinese ambassador revealed as much when he said that China would be "uncooperative in all multi-and bilateral arms control fora if the US proceeds with NMD".[34] That this cannot be taken lightly stands vindicated by the fact that for the Chinese leadership, the entire argument of non-proliferation is also more or less linked to their concept of "asymmetric warfare". China upholds that if one has to fight a technically more advanced and financially stronger adversary, then it would be foolish to do so through frontal confrontation. Deception, instead would be a better alternative, as argued by Sun Tzu who said, "All warfare is based on deception. Therefore, when capable of attacking, feign incapacity; when active in moving troops, feign inactivity. When near the enemy, make it seem that you are far away; when far away, make it seem that you are near."[35] China has internalised this thought process and amalgamated the same into its military thinking.

While this should be of concern to the international community in general, the matter acquires greater significance for India given the close Sino-Pak nuclear and missile nexus. In any case, Beijing is acknowledged to have a long history of selling nuclear and ballistic missile technology to other countries. But, with reference to the NMD, China has not hesitated to openly state that in a bid to restore the strategic balance disturbed by the NMD, it shall be compelled to reconsider its approach and subscription to all arms control and non-proliferation agreements.

Proliferation could then well be on the anvil. Apart from the purely strategic reasons, commercial considerations may also dictate such a policy. The development of sophisticated counter-measures or an NMD resistant system would call for a substantial influx of finances that would impose a strain on the country's economy. In order to recover some of these costs, China could see economic (as well as strategic) sense in selling the relatively older technologies to other countries. In any case, China has never let any international agreement stand in the way of such transfers. If it so desires in the future, the NMD, in fact, shall provide it with a perfect excuse to proliferate with impunity. As Brahma Chellaney writes, citing an intelligence report from the Pentagon, "Beijing has found ingenious ways to proliferate *on a consistent basis without technically breaking agreements with the (United States).*"[36]

On another front, the Chinese strategy of dealing with the NMD has involved criticising it at every international forum for its potential impact on global security and stability. The possibility of an upgradation of the NMD to facilitate an eventual deployment of space-based assets has caused maximum concern to China. Consequently, Chinese spokespersons have elaborately spelt out the dangers from militarisation of outer space at every international and bilateral discussion. In fact, negotiations for the conclusion of an FMCT remain stalled owing to China's insistence that an ad hoc committee be instituted to simultaneously consider the issue of PAROS. China could also be keeping its option open with regard to accumulating more fissile material in case the NMD requires it to increase its arsenal. Similarly, the need to develop counter-measures such as manoeuvring warheads could make Beijing opt out of the CTBT and conduct the necessary tests.[37]

The US may not feel the need to be unduly concerned with these developments in China since even a ten or twenty-fold increase in the Chinese ICBM force would not alter the strategic balance with Washington in Beijing's favour. But, from the Indian point of view, any increase in Chinese missiles, even if of limited or no value for Washington, would however, carry ominous portents for the region. In any case, China has never hidden its intention or desire to play a more dominant role in Asia, and the upgradation of its missile force shall only add to this capability. At this point of time, the nuclear balance is tilted in favour of China, even as India is struggling to build a credible minimum deterrent. China's nuclear modernisation plans that seek to replace its ageing, liquid-fuelled, cave and silo-based missiles with newer solid-fuelled, mobile missiles with shorter launch preparation period, or attempts to MIRV the existing arsenal, or to develop counter-measures can only add to India's insecurities, at least until such time as the neighbours resolve their border differences and embark upon meaningful nuclear confidence building measures.

In any case, the Chinese are engaged in a process of military modernisation. While some argue that this process is going to run its course independent of the US plans on the NMD,[38] an important difference must be kept in mind. Had the Chinese been engaged in the modernisation of their conventional and nuclear forces at a time when the international security environment had been one of relative stability, it would have been perceived as an aggressive act, certain to invite criticism from others. But, with the Chinese activity taking place at a time when the prospect of NMD deployment looms large, the former appears to have gained a measure of legitimacy. It is as if the NMD has provided a useful shield to Beijing to carry on with its military modernisation, a development that India's security establishment shall have to take due note of.

Impact on Nuclear Non-Proliferation and Disarmament

President Bush claims, "Defences can strengthen deterrence by reducing the incentive for proliferation." The logic that underlines this argument is that since enemy missiles shall not be able to penetrate the missile shield, countries, over time, shall sense increasingly less utility in such weapons. Indeed, the argument appears acceptable on face value, but certain factors must be considered before it can be established that erecting defences would actually be the most viable way of devaluing a lethal WMD.

First, there is an increasing gap between the rate at which the strategic threats and the missile defences are developing. According to the Rumsfeld Commission, the threats from North Korea could materialise as early as within the next five years, and those from Iran and Iraq before the end of this decade. In contrast, a limited missile defence capable of guarding against the North Korean threat is unlikely to be in place before 2007-08 and protection against missiles from the Middle East not before the end of the next decade. The time taken in developing viable BMD systems would also be utilised by states aspiring for these missiles to improve and enhance their penetration capabilities. Hence, any BMD system that emerges may be partly outdated and might never be able to guarantee absolute security against incoming missiles.

Second, in the context of the NMD, the US primarily treats ICBMs as constituting risks to its security. For small states, building ICBMs is certainly an economic and technological challenge and an effective missile defence could indeed strip them of their value and the incentive to develop them. However, for most countries that undertake to acquire/develop WMD and their delivery vehicles, the rationale mostly exists within the region. Power projection, deterrence and safeguarding national sovereignty against blackmail and coercion in the regional context are the primary reasons that prompt them to go down this path. In such cases, where the area of operation is to be limited, missiles with short, medium, or intermediate ranges should more than suffice and these are relatively easier and cheaper to build or acquire than ICBMs. While these may not be of immediate concern to the US mainland, for countries within range of such missiles, the situation becomes destabilised. Though a TMD is envisaged as providing protection against such threats, it is not likely to be available too soon in the near future. In the interim then, regional states apprehending a threat to their security could find it prudent to build a WMD missile deterrent capability of their own.

Third, despite all the hyperbole over the BMD, there can be no foolproof guarantee that every incoming missile shall be intercepted in time and shot down. This risk is further compounded by the ongoing research on developing counter-measures. In fact, way back in 1961, Eric Burgess had written:

> It certainly cannot be concluded that an attacker will merely use simple warheads, letting his ballistic missiles perform like high-altitude research vehicles. We must expect that the warhead will be protected by counter-measures against the AMM (anti-missile missiles), including decoys, missiles launched in front of the actual ICBM, and expandable radar reflectors ejected from the ICBM afterbody or from the reentry body itself...The reentry body might itself be supercooled by refrigerants before reentry to upset the infrared detectors.[39]

A similar word of caution was sounded by the National Intelligence Estimate in September 1999 when it reported:

> We assess countries developing ballistic missiles would also develop various responses to US theater and national defences. Russia and China each have developed numerous countermeasures and probably are willing to sell the requisite technologies.
>
> Many countries, such as North Korea, Iran, and Iraq probably would rely initially on readily available technology—including separating reentry vehicles (RVs), spin-stabilised RVs, RV reorientation, radar absorbing material (RAM), booster fragmentation, low-power jammers, chaff, and simple (balloon) decoys—to develop penetration aids and countermeasures.[40]

Hence, the risk of penetration into a BMD might prompt smaller countries that face missiles threats from within the region to see better sense in developing a WMD retaliatory capability rather than wasting scarce resources on an elaborate and expensive missile defence system that might still be penetrable. Proliferation, therefore, might not only be merely arrested with the erection of a missile defence system, but may actually be encouraged.

Fourth, the current focus on missile defences notwithstanding, traditional wisdom as well as empirical observations of the Cold War era suggest that the only effective way of deterring WMD threats is by the acquisition of a similar capability by the adversary as well. The balance of terror then establishes a sort of stability and peace, however, uneasy and precarious. Increasing awareness of the susceptibility of BMD to counter-measures and the spiralling costs of BMD research, development and deployment may most likely reinforce faith in deterrence, especially for smaller states unable to afford an expensive BMD system. This, in turn, would signify nuclear and missile proliferation. In fact, the prospects of proliferation are brightened due to the relative ease in accessibility to nuclear material, technology, expertise or even the weapon itself owing to the wide-scale dismantlement of several categories of nuclear weapons in both the US and Russia. This has already resulted in a surplus of weapons usable material. Further, given the lack of adequate storage and safety requirements in Russia, cases of smuggling of fissile material, though in miniscule amounts, have been reported. The political and economic upheaval that accompanied the fracture of the USSR and the collapse of centralised Soviet control has multiplied the chances of illicit nuclear commerce developing as a viable economic activity, especially if more and more countries begin to see merit in acquiring a nuclear weapons capability.

Fifth, NMD deployment by the US has led it to withdraw from the

ABM Treaty. This withdrawal itself has set a precedent that would impact on the fate of international treaties, especially those relating to non-proliferation and arms control. If a treaty like the ABM that was widely believed to be the cornerstone of all arms control agreements[41] could be dispensed with because of the changes in the perception of the strategic context, then a case can be made for amending or abrogating the NPT as well owing to the changes in the international arena after Pokhran II. The US' unilateral withdrawal from the ABM Treaty even in the face of opposition from so many countries, including the European allies, would directly impinge upon the sanctity of such international agreements. Other countries may be tempted to do the same when it no longer suits their interest to abide by a treaty arrangement. Until now there is no example of a country walking out of its commitments agreed to under a treaty. But the US' decision to step out of a treaty that does not even have a withdrawal clause could prompt others for exercising the same option with regard to the NPT or the CTBT that do have a provision for withdrawal.

Hence, the overall impact of the NMD on global non-proliferation trends does not appear to be a reassuring one. By the time BMD systems are perfected to meet the threats perceived today, the threats themselves could have altered in nature, gravity or complexion. Of course, it is not the intention of this chapter to undermine or underestimate the possible technological breakthroughs that might be achieved in the coming years, but to question whether trudging down this path would not lead to the making of new threats. Suspicious attitudes and actions premised on mistrust can naturally be expected to breed more of the same. American threat perceptions that are prompting it to build a missile shield could end up upsetting existing international security arrangements and fostering a worldwide lack of confidence. And yet, despite causing so much turbulence, the shield may not act with a credible guarantee of the possibility of every missile being neutralised well in time.

At another level, President Bush has held out a sop of sorts to the campaigners of disarmament by announcing that he would follow up NMD deployment with a change in the size, composition and character of US nuclear forces. While this would have been a laudable objective by itself, coupled as it is with the NMD, it loses much of its value since the reduction in the US nuclear arsenal is unlikely to be seen as a case of "lead(ing) by example" by the others. Neither Russia nor China is likely to be impressed by these reductions. These cutbacks are rather representative of a rationalisation of the US nuclear forces and not part of any strategy that envisages the elimination of nuclear weapons over time. With a protective missile shield in place, there would be little need for the US to maintain the huge stockpiles that now exist. The same, however, may not hold valid for Russia and China. As has been discussed in an earlier section of this chapter, China, in the wake of the NMD, is more prone to feel the need to refurbish and expand its nuclear forces in order to restore a semblance of strategic stability vis-à-vis the US. Russia too, though presently constrained by its economic conditions to maintain even the present level of its nuclear forces, may feel compelled by the NMD to retain a larger stockpile by going back on its committed reductions under the START II and even re-MIRVing its existing missiles. In fact, President Putin has hinted that doing so would rather allow Russia to "maintain its deterrent more economically." Moreover, any turnaround in the economy, whenever it occurs, could tempt Russia to further consolidate its nuclear arsenal especially in the wake of the deterioration in its conventional military might. The collapse of the START Treaties would also end mutual verification of future arms reductions and undermine the incentives for doing so.

Meanwhile, US reduction of its nuclear weapons cannot mean anything significant from the nuclear disarmament perspective. It is a move designed purely on the basis of national interest and cannot contribute to international stability in any significant way, at least not for Indian security. Rather, the very fact that the president should have spoken of the nuclear reduction in the same breath as "nuclear weapons still have a vital role to play in our security and that of our allies" negates any positive signal that it could have been meant to send. How and why should it be expected that other countries would be ready to acquiesce to the existence of these WMD in the hands of a few and also suffer silently their going under a defence shield without wanting to acquire the same themselves? While the erection of a defence system might be a technologically and economically challenging prospect, acquiring the sword would be much easier, especially so in the present times. Therefore, by raising insecurities, the BMD systems may instead trigger an increase in the global count of nuclear weapons and their delivery mechanisms.

Conclusion

The US missile defence initiative encompasses a number of moves. For instance, it seeks to build a cooperative world order premised at some time in the future on "joint defence" or collective security. It envisages a move away from the Cold War logic of nuclear deterrence. It embodies a reduction in US nuclear forces. As individual moves, these do not necessarily have negative overtones.

However, when considered holistically as a package, most of these initiatives lose their positive sheen. The cooperative world order is to be formalised (and hence polarised) between those whom *the US* considers as its friends and allies and those that it does not. Actions against or in favour of them shall be controlled by the US, thereby reinforcing unipolarity. Meanwhile, with an overtly established policy of preemptive use of military force, the US has ascribed to itself immense powers of retribution. Also, the move away from nuclear deterrence envisages a shift towards missile defence and not the elimination of nuclear weapons. In fact, to view the proposed nuclear reductions from the perspective of nuclear disarmament would amount to missing the wood for the trees. On the whole, therefore, the US security strategy may actually exacerbate insecurities and prove detrimental to international stability in ways more insidious than are presently being considered.

The implications of the BMD are many. It will definitely set into motion an action-reaction process, the ripples of which are not likely to leave Indian shores untouched. Therefore, any cooperation—material or conceptual—that New Delhi agrees to extend to the US plan must be carefully weighed against the costs—tangible and intangible—that might accrue to it in the process of helping the US deal with its threat perceptions.

Through the NMD, the US is certainly seeking freedom of military action. As President Nixon had said even at the time of the ABM Treaty, "If you have a shield, it is easier to use the sword."[42] Meanwhile, the US is also investing its resources in sharpening the sword through constant research and development in its offensive military capabilities. Superiority in offence and defence would most likely give a boost to unilateralist tendencies. In doing so, however, the US is exhibiting insensitivity to the concerns or interests of other countries—a perception that could foster a confrontational world order. Such a scenario shall be counter-productive for India and force it to invest undue resources in activities other than economic development that must be its present priority.

Therefore, it must be reiterated that for India, as also for the rest of the world, a guarantee of absolute security lies not in erecting missile defences that could still be penetrated, but in doing away with the threat of nuclear weapons and their delivery mechanisms *in toto*. As the French President Jacques Chirac has warned,

> If you look at world history, ever since men began waging war, you will see that there's a permanent race between sword and shield. The sword always wins. The more improvements are made in the shield, the more improvements are made to the sword. We think that with these systems we are just going to spur sword makers to intensify their efforts.[43]

Missile Disarmament

An effective end to this protracted cycle can only be found in measures that devalue and delegitimise these WMD and their delivery vehicles, thereby reducing their role and acceptability in inter-state relations. The true way out would lie in helping engender an international climate in which countries become less dependent on their military prowess and, therefore, less inclined to use missiles in the first place. Unfortunately, though, the issue of missile disarmament has been largely ignored in international arms control and disarmament negotiations. This, in fact, was recently pointed out by UN Under-Secretary-General for Disarmament Affairs Jayantha Dhanapala: "Why is public debate mired today in a duel between deterrence and defence, with scant attention to missile disarmament?"[44]

Undoubtedly, as the dangers of a race between offence-defence weaponry become imminent, the need for an international initiative to control ballistic missiles becomes even more urgent. While the MTCR has now existed for several decades, that it has not been effective enough is proved by the fact that the US is today feeling the need for missile defence. From missile capability being the exclusive preserve of only seven countries when the MTCR came into being, today nearly thirty countries are believed to have the capability to build these delivery mechanisms. Obviously, then, the MTCR has not been able to check missile proliferation and there is a need to strengthen international ballistic missile controls through the adoption and evolution of some other measures. Suggestions to this effect range from establishing a Code of Conduct, bolstering export controls, missile monitoring and launch-notification agreements to the conclusion of a more far-reaching disarmament treaty establishing a global missile ban.

Even though the prospects for such a comprehensive disarmament regime based on multilateral agreements currently seem remote, especially in view of the attitude of the US Administration to multilateral arms control, certain intermediate options have been suggested in placing restrictions on missile testing, and the creation of missile-free zones. Test restrictions would effectively prevent new missile designs and limit modification of traditional technology, although unsophisticated indigenous missile systems could still be developed and deployed with minimal testing. A ballistic missile test-flight ban would preclude the testing of new missiles and reduce the chances of accidental or unintentional war. Therefore, a moratorium on the further development, testing and deployment of ballistic missiles or a "missile freeze" could serve as a temporary measure while countries negotiate missile disarmament.

Models for missile disarmament, in fact, already exist in some forms. For instance, in 1992, following up on a proposal discussed between then Presidents Ronald Reagan and Mikhail Gorbachev at the 1986 Reykjavik Summit, the Federation of American Scientists (FAS) had developed a model (a complete draft treaty, in fact) for the elimination of ballistic missiles entitled Zero Ballistic Missiles (ZBM).[45] This envisaged the total elimination of offensive ballistic missiles through unilateral declarations and regional and global multilateral agreements. The ZBM proposal was based on a step-by-step incremental approach that would start with bilateral cuts between the USA and Russia, creation of ballistic missile-free zones, the convening of an international Missile Conference, the institution of an International Agency for Ballistic Missile Disarmament, and finally an agreement on the varying schedules necessary to reach zero ballistic missile capability.

However utopian these measures may seem in the present context, there can be no denying the fact that just as in the case of nuclear disarmament, missile disarmament too can be the only effective and sustainable method of obviating the dangers of proliferation and use. Would such a ban be in India's favour? Yet again, the answer to this question must draw upon India's experience and commitment to nuclear disarmament. As has been stated earlier in the chapter, India faces tremendous missile threats from its neighbours in the east and the west. Geographical proximity enables Pakistani and Chinese weaponry to reach India with no more than short and medium range ballistic missiles, technology for which is comparatively easier to develop, or even obtain from others, as Islamabad has well demonstrated. In case a *universally applicable, non-discriminatory ban on missiles* can be established, it would certainly go a long way in reducing India's threat perceptions.

Of course, critics of this proposal could raise the same arguments against missile disarmament as are put forth against nuclear disarmament. Indeed, the knowledge to build missiles would always remain with countries, and cheating could destabilise international security. However, risks of this kind could be minimised through effective and universally applicable means of verification. National or international technical means of verification could focus on observable rocket characteristics (number, size, range, payload, deployment mode, launch preparations, flight trajectory) that provide indications of rocket type and performance. Most of the infrastructure of a missile programme such as production facilities, test ranges, tracking and communication facilities, missile containers and missile-carrying vehicles, etc. are fortunately highly visible and, hence, could lend themselves to easy monitoring. Remote sensing in the visible, infrared or radar spectra, based on satellites, aircraft or on the ground allows observation of missiles and the related launch and test facilities. Some verification tasks could be performed by commercial satellites too.

The biggest complication, however, could arise in the difficulty to differentiate between the dual-use of ballistic missiles and satellite launch vehicles (SLV). With the technology being easily convertible, distinguishing between both rocket types would pose a problem. However, some functional differences and operational characteristics such as in the basing mode, testing procedures, payload, flight trajectory, guidance systems and re-entry could still be used to improve detection.[46] Also, another system of safeguards for space launchers could be negotiated to place some of the "most critical" items under supervision by an international organisation. International cooperation in civilian space programmes would nevertheless be important for containing the use of space technology for missile development.

Technical means of verification could be further complemented with inspections. This, of course, would call for cooperation and a willingness on the part of states to allow unimpeded and speedy access to suspect sites and to exchange information. Systematic inspections of all ballistic missile-related sites to provide basic information on an initial balance, could then be followed up with random short-notice inspections of declared sites as well as challenge inspections of undeclared sites.

Although these proposals appear unrealistic today, they are going to become increasingly more, and not less, important as a means of preserving stability and reducing uncertainty in a world of dangerous and costly missile defences. Some of this understanding was reflected in the negotiations held on June 17-19, 2002, between delegations from nearly 100 countries in Madrid to draft an International Code of Conduct (ICOC) against ballistic missile proliferation. The ICOC, which would only be politically binding and not an article of international law, seeks to address some of the shortcomings of the MTCR by introducing demand-side controls on the acquisition of ballistic missile technology. The Madrid conference built upon previous discussions held in Paris in February, and included many ballistic missile capable states including Russia, China, India, Israel, Pakistan, the United States, and all EU member states. Syria and North Korea chose not to attend either the Paris or Madrid meetings, while Iraq was not invited. Iran—an active participant in the Paris talks—pulled out of the Madrid meeting at the last minute. Interestingly, the ICOC proposes a number of confidence building measures such as provisions for annual declarations on ballistic missile and SLV activities and policies, invitations to host international observers at space vehicle launches, etc. These are aimed at promoting transparency and enhancing stability. However, critics of this initiative are many. It is nevertheless possible that an ICOC against ballistic missile proliferation would emerge as a complementary tool to other non-proliferation efforts. This would represent an important step towards a more inclusive approach to engagement and agreement on missile proliferation. Therefore, the importance of the ICOC as a multilateral initiative should not be overlooked and its universal acceptance should be encouraged.

Notes

1. The BMD is like an umbrella concept that encompasses both the NMD and the TMD. While the former aims at protecting the continental US from a limited ICBM attack, the latter seeks to provide protection to forward deployed troops and naval fleets against short range, medium range and intermediate range ballistic missiles. The Bush Administration, however, no longer makes any clear distinction between the NMD and the TMD. As the US secretary of defence has said, "...I have concluded that 'national' and 'theater' are words that are not useful...What's national depends on where you live, and what's theater depends on where you live," Mark Bromley, "European Missile Defence: New Emphasis, New Roles", BASIC Paper no. 36, May 2001, http:///www.basic.intorg

2. North Korea, Iran, Iraq and Libya are generally referred to as the states of proliferation concern by the USA.

3. President Bush's speech on missile defence, delivered at the NDU, Washington, May 1, 2001 and as available on http://www.acronym.org.uk

4. "US Missile Defence Director Cites Progress, Challenges", *Washington File,* US State Department, March 12, 2001. This article is adapted from a speech by Ronald Kamish to the Greater Fairbanks Chamber of Commerce, Alaska, on March 2, 2001 and quoted in *Disarmament Diplomacy,* no. 55, March 2001.

5. As quoted in Thomas Etzold and John Lewis Gaddis, eds., *Containment: Documents on American Policy and Strategy, 1945-1950* (New York: Columbia University Press, 1978).

6. Soon after his speech at the NDU, President Bush dispatched senior officials from his Administration to the capitals of major countries, including Moscow, Beijing, London and New Delhi to brief them on the proposed NMD initiative. However, it has since been disclosed that the whirlwind tour of the US representatives did not bring any positive tidings for the US plan of creating a new framework for strategic thinking on nuclear deterrence, See *Proliferation News* at http://ceip.org

7. Vladimir Radyuhin, "Russia 'Sways' India on ABM Treaty," *The Herald,* June 7, 2001, p. 11.

8. President Bush's speech, n. 3.

9. Ibid.

10. *Executive Summary of the Report of the Commission to Assess the Ballistic Missile Threat to the United States,* July 15, 1988, p. 1. As available on http://www.fas.org/irp/threat/bm-threat.htm It may be recalled that during the mid-1990s, the US National Intelligence Estimate (NIE) had downplayed any new missile threat to the country. In November 1995, the NIE had noted that North Korea was unlikely to obtain the technological capability to develop a long range ICBM that could threaten the US heartland, and that there was an even less chance of Iran or Iraq developing such a capability. The Rumsfeld Commission trashed this premise.

11. As of now, North Korea's nuclear programme continues. The Clinton Administration had been able to dissuade Pyongyang from testing the Taepo Dong 2. In return for the lifting of some US economic sanctions, North Korea had agreed to halt its missile tests in September 1999 at least till such time as "the talks (with the US) are underway." See Kathryn Tolbert, "North Korea Confirms it Will Not Test Missile", *Washington Post,* September 25, 1999. In October 2002, North Korea once again startled the world by acknowledging that it continued to pursue a nuclear weapon programme. "North Korea Has Two Nuke Bombs: US," *The Times of India,* October 19, 2002.

12. n. 3.

13. Ibid.

14. Indeed, it was one such assessment of the threats to its security that had led India to conduct nuclear tests in May 1998 after 25 years of nuclear restraint.

15. Public Law 106-38, July 22, 1999. As quoted in Ivo H. Daalder, James M. Goldgeier and James M. Lindsay, "Deploying NMD: Not Whether, But How," *Survival,* vol. 42, no. 1, Spring 2000, p. 10.

16. n. 3. Emphasis added.

17. For more on this, see the chapter on the CTBT in this book.

18. Details on the state of this treaty can be found in the chapter on the FMCT in this book.

19. Henry D. Sokoloski, "Counter-Proliferation," *Bulletin of Atomic Scientists,* March-April 2001, p. 63.

20. The idea that the "rejection of the nuclear shield made the nuclear sword less dangerous" has echoed in several writings. For a more recent articulation, see Igor Ivanov, "The Missile Defence Mistake", *Foreign Affairs,* vol. 79, no. 5, September/October 2000.

21. For more on both these viewpoints, see Keith B. Payne, "Looming Security Threats: The Case for NMD," *Orbis,* Spring 2000, p. 189.

22. Report of the Commission to Assess the Ballistic Missile Threat to the US, Executive Summary, July 15, 1998.

23. Some American scientists have cynically observed that "the probability of the NMD coming into force in the next few years is only moderately higher than a snowflake's chance in hell." V. Sudarshan and Yukteshwar Kumar, "The Triangle," *Outlook,* May 21, 2001, p. 44.

24. Kamish, n. 4.

25. Ibid.

26. NBC news report, June 2000. North Korea has now been established as another source of Pakistani nuclear armoury. Engaging in sensitive trade, North Korea provided missiles to Pakistan in exchange for nuclear weapons. "Pak, N. Korea Bartered Missiles for Nukes," *The Times of India,* October 19, 2002.

27. These considerations though appeared to weigh little for a mischievous Pakistani report that claimed in August 2001 that India was all set to sign a deal to purchase six S-300 anti-ballistic missile defence systems and build indigenously 19 more similar systems over the next 15 years. "India May Get ABM Defence System," *The Dawn,* August 2, 2001.

28. For an interesting Russian perspective on the NMD, see Ivan Safrunchuk, "Russian Views on Missile Defences," *Pugwash Occasional Papers,* vol. 2, no. 2, March 2001, pp. 32-43.

29. For a comprehensive Chinese critique of the NMD, see transcript of Ambassador Sha Zukang's Press briefing in Beijing on March 14, 2001, as carried in *Disarmament Diplomacy*, issue no. 55, March 2001.

30. Chinese officials refuse to provide any figures of ICBMs and themselves quote American sources to say that China is believed to have between 18 and 24 such missiles!

31. Official Chinese statements as cited in John Lewis and Xue Litai, *China Builds the Pomb* (Stanford: Stanford University Press, 1988), pp. 241-242 and quoted by Michael McDevitt, "Beijing's Bind," *Washington Quarterly*, Summer 2000, p. 181.

32. A *New York Times* report cited by Charles D. Ferguson and John E. Pike, "Will Missile Defence Shoot Down Arms Control?" *Yaderny Kontrol (Nuclear Digest)*, vol. 5, no. 4, Fall 2000.

33. McDevitt, n. 31.

34. Ambassador Hu Xiaodi's statement at the CD as quoted by Sverre Lodgaard, "European Views of the US NMD programme," *Pugwash Occasional Papers*, vol. 2, no. 2, March 2001, p. 50.

35. Sun Tzu, *The Art of War* by General Tao Hanzhang, Wordsworth Reference, (Denmark, 1995), p. 101.

36. Brahma Chellaney, "New Delhi's Dilemma," *Washington Quarterly*, Summer 2000, p. 149.

37. Li Bin, "Impact of US NMD on Chinese Nuclear Modernisation," *Pugwash Occasional Papers*, vol. 2, no. 3, June 2001, p. 67.

38. See Brad Roberts, "US Ballistic Missile Defence: Implications for Asia," in Jasjit Singh, ed., *Reshaping Asian Security* (New Delhi: IDSA, 2001).

39. Eric Burgess, *Long-Range Ballistic Missiles* as cited in Richard L. Garwin, "A Defence that Will Not Defend", *Washington Quarterly*, Summer 2000, pp. 115-116.

40. Central Intelligence Agency, National Intelligence Estimate, "Foreign Missile Developments and the Ballistic Missile Threat to the United States Through 2015". http://www.cia.gov/cia/publications/nie/nie99msl/html

41. Russian officials had counted upto more than 30 strategic accords as being tied to the continuation of the ABM Treaty. More on the European opposition to the US withdrawal from the ABM Treaty is available in Wade Boese, "Bush Meets Opposition to Missile Defence While in Europe", *Arms Control Today*, July-August 2001.

42. Lodgaard, n. 34, p. 52.

43. *New York Times*, December 17, 1999.

44. "Eliminating Nuclear Arsenals: The NPT Pledge and What It Means," speech by Jayantha Dhanapala to the All-Party Group on Global Security and Non-

Proliferation, House of Commons, London, July 3, 2000. For full text, see *Disarmament Diplomacy,* no. 47, June 2000.

45. "Revisiting Zero Ballistic Missiles—Reagan's Forgotten Dream", FAS Public Interest Report, May/June 1992; L. Lumpe, "Zero Ballistic Missiles and the Third World," *Arms Control Today,* vol. 14, n. 1, April 1993, pp. 218-223; A. Frye, "Zero Ballistic Missiles", *Foreign Policy,* no. 88, Fall 1992, pp. 12-17.

46. J. Scheffran, "Dual Use of Missiles and Space Technologies," in: G. Neuneck, O. Ischebeck, eds., *Missile Technologies, Proliferation and Concepts for Arms Control,* (Baden-Baden: Nomos, 1993), pp. 49-68.

7

Weaponisation of Outer Space

Manpreet Sethi

The demise of the treaty between Russia and the USA on the Limitation of Anti-Ballistic Missile systems (ABM) came about quietly in the month of June 2002 on the completion of the period of notice for withdrawal rendered by Washington in December 2001. Russia, some countries of Europe, and China expressed no more than murmurs of regret on the occasion and the turbulence in strategic stability that had been initially predicted was noticeably absent. US diplomacy had, indeed, utilised well the six months of the notice period to either win over countries to its position on the NMD and the abrogation of the ABM, or made them resign to accepting the inevitable.

Whatever be the long-term response of other countries to the dismantling of the ABM, and whatever be the nature of international relations and strategic stability in the future, one disquieting aspect of the BMD that demands immediate and urgent attention is its implications for weaponisation of outer space.[1] This would include the deployment of weapons in space, including those of mass destruction, that have the capability of attacking objects in space, air or on the ground. It would also include weapons against space objects deployed in space, air or on the ground. The repertoire of space weapons is large and varied. They could range from nuclear explosions that produce electromagnetic pulse or increased charged particle radiation; kinetic energy weapons such as missiles, kill vehicles and collision devices to hit space objects; directed energy weapons that can emit microwaves, laser beams or atomic particles; and, miniature satellites that can dock onto another one to cause malfunctioning or manipulation.

The 1972 ABM Treaty had placed a check on the weaponisation of space by forbidding the countries from developing, testing or deploying ABM systems. This also ruled out the deployment of interceptors or sensors in space that could substitute ground-based ABM radars. It had also banned ASAT (anti-satellite) systems with ABM capabilities, besides outlawing concealment from, and interference with, the other side's "national technical means of verification (NTM)",[2] so crucial for upholding the credibility and confidence in arms control treaties.

The US withdrawal from the ABM Treaty leaves the 1967 Treaty on Principles Governing the Activities of States in the Exploration and Use of Outer Space, including the Moon and other Celestial Bodies, or what is commonly referred to as the Outer Space Treaty (OST) as the only legal instrument prohibiting some degree of space weaponisation.[3] Defining outer space as a common heritage for peaceful purposes, with no boundaries, no private property for exploitation, and no sovereign rights, the treaty bans testing of weapons, conduct of military manoeuvres or the establishment of military bases in space, on the moon or other celestial bodies. But it does accept the passive military use of space for reconnaissance satellites, surveillance and early warning systems as well as communications. It also does not prohibit other conventional weapon systems or ASAT weapons from being used in or from space. Neither does it cover the transit of nuclear weapons through space or their being launched from earth into space for destroying incoming missiles.

Of course, at the time when the OST was concluded, the possibilities of deployment of weapons in outer space, or even their use in the medium, were subjects for science fiction. Today, though, advancements in a vast range of space technologies make the possibility of space-based warfare appear not so removed from reality. Consequently, the gap between present day capabilities and the legal instruments once concluded has widened. With no treaty specifically addressing the issue of weaponisation of outer space, and with the present US Administration showing a resolute determination to move ahead with strategies aimed not just at *protecting* its substantive military and commercial assets in outer space, but at *dominating and controlling* activities on earth from space, the risk of an arms race, or even conflict, in space cannot be discounted.

In the light of this reality, this chapter argues that attempts to dominate space militarily could seriously backfire, lead to a new space arms race and increase the vulnerability of important commercial, communication, verification and intelligence assets in space. It traces the direction of present space-related technical developments and analyses the trends that are likely to prevail in the policies of countries capable of placing weapons in space. It highlights the possible difficulties that could arise in exercising arms control in the extra-terrestrial realm, particularly given the thin line dividing commercial and military use of space. Further, considering that the CD remains deadlocked, with no realistic hope of any kind of substantive negotiations on PAROS,[4] the chapter explores other possibilities to regulate the non-aggressive, non-offensive and peaceful uses of outer space. It also examines the Indian position on these issues.

The Relationship Between BMD and Weaponisation of Space

As has been discussed in the previous chapter, the US is engaged in the process of testing and building a multi-tiered missile defence system that is capable of intercepting and annihilating intermediate range or inter-continental ballistic missiles. This has been criticised by Russia and China, besides America's allies in Europe as well, for three main reasons: destabilising relations between nations by undermining deterrence; discrediting treaty-based collective regimes of non-proliferation and security; and leading to strike weapons being placed in space. It is this third issue that is causing concern to several countries as they espy a direct link between the NMD and a phenomenal enhancement of US space capabilities. In fact, it has been established that a land or sea-based missile defence system capable of intercepting missiles would automatically be capable also of intercepting satellites in low earth orbit (LEO).[5] And, the maximum number of earth-orbiting satellites for military and civil purposes is concentrated in LEO. Shooting down a satellite here is believed to be easier than shooting down an ICBM in mid-course, since the exact trajectory of the latter is not known, and also because of the counter-measures used. The satellite in orbit, meanwhile, follows a known path and can easily and directly be targetted by a missile defence system. This raises the possibility of the BMD directly taking war into outer space. At the same time, indirectly too, by contributing to the military infrastructure in space for communication, reconnaissance, surveillance and tracking, the BMD would increase the likelihood of conflict in space. Whether land-based or sea-based, missile defence would require communication and reconnaissance assets to be placed in space. In fact, the Space-Based Infra-Red Satellites (SBIRS) that will comprise a surveillance and tracking system would be part of the larger overall network of space assets of the NMD.

But US officials are hesitant to acknowledge the linkage between BMD and space weaponisation. Rather, in an attempt to calm international concerns, they stress the "limited and essentially defensive nature" of the NMD designs. Paul Wolfowitz, US deputy defence secretary, has emphasised, "It is not an effort to build an impenetrable shield around the US. This is not Star Wars. We have a much more limited objective to deploy effective defences against limited missile attack."[6]

Nevertheless, several official US pronouncements do reveal a correlation between BMD and space-based weapons (SBW). In fact, US Secretary of Defence Donald Rumsfeld has himself made the connection several times. It is well known that Rumsfeld considers space to be of tremendous importance in future US military strategy. Even before he took over as secretary of defence in the present US Administration of President Bush, he had chaired the Commission to Assess United States National Security Space Management and Organisation in 2001, wherein he had recommended that US national space policy be brought into the *centre of defence planning, encompassing an early review and revision of policy priorities with the objective of developing concepts for operations and capabilities for space, including weapon systems.*[7] Subsequently, he has identified *"maintaining unhindered access to space and protecting US space assets from enemy attack"* as one of the "transformational goals" before the US military.[8]

Raising the spectre of a "Space Pearl Harbour," the Rumsfeld Commission had argued that US military capabilities would need to be enhanced to enable:

- Assured access to space and on-orbit operations.
- Space situational awareness.
- Earth surveillance from space.
- Global command, control and communications in space.
- Defence in space.
- Homeland defence.
- Power projection in, from and through space.

It had further recommended strengthening intelligence capabilities, creation of a cadre of space professionals, and restructuring of the decision-making to bring national security space policy into the mainstream and under the "deliberate leadership" of the US president. The Commission had argued that the US should participate actively in shaping the space legal and regulatory environment. In fact, it had cautioned the US against *"agreements intended for one purpose that, when added to a larger web of treaties or regulations, may have the unintended consequences of restricting future activities."* It is not surprising, therefore, that US officials in the CD have consistently refused to move towards negotiation of an agreement on limiting military activities in outer space. Rather, they propagate the viewpoint that "the current international regime regulating the use of space meets all our purposes. We see no need for new agreements."[9]

This thinking reflects the tremendous significance that US officials attribute to space in the coming years. Several service documents, the Quadrennial Defence Review and the Nuclear Posture Review among others, designate space as "a fundamentally new and better way to apply military force—by promptly striking adversary centres of gravity, or minimising or bypassing high cost, high risk conflicts".[10] Current US space-related military strategy draws upon four operational concepts as developed by the US Space Command over the last few years. These include:

- Control over, or superiority in, space or force application from space. It envisages the ability to assure or deny access to, and freedom of, operations within space.
- Global engagement that combines integrated, focussed surveillance and missile defences with a potential ability to apply force from space. It calls for "full spectrum dominance of the battle-space" by the deployment of a space-based anti-ballistic missile system, land-based and sea-based anti-satellite systems, and a variety of orbiting systems that can strike terrestrial targets. The NMD is designated as a "key" to "global engagement capabilities." Noting the rationale for the NMD and global engagement, US Space Command Vision 2020 looks to "[I]ntegrating Space Forces into war-fighting capabilities across the full spectrum of conflict".[11]
- Full force integration of space forces and space-derived information for air, land and sea operations through the application of precision force from, to and through space. The core of this strategy is to

weaponise space to enable the US to fight a war in all four media to protect American interests and investments.

- Global partnerships for augmenting military space capabilities through exploitation of civil, commercial and international space systems, including bilateral partnerships with US allies in Europe and Asia.

Given the above, it is not surprising that the US military is conducting war-games envisaging such scenarios. One such war-game that focussed on space as the primary theatre of operations, rather than just a supporting arena for combat on earth, was played out at the Schriever Air Force Base in Colorado.[12] These scenarios draw sustenance from positions as advanced by General Ralph Eberhart, United States Air Force Commander in Chief, North American Aerospace Defence Command and United States Space Command, who has argued in his Congressional testimony that:

> It is time to push up the "space superiority throttle." We have left this throttle at a low power setting for too long. We must ensure our continued access to space, to deny space to others when directed...This is a medium crucial to our American military operations and one we'll have to fight for in the future.[13]

In his role as Secretary of Defence, Donald Rumsfeld has elevated the importance of space in US military strategy and a four-starred air force general now directs the military space programmes. In May 2001, in a significant move, space activities were consolidated within the administration of the Pentagon with the air force being designated as the "executive agent for space". In fact, the air force chief of staff has made it amply explicit that, "...our goal is to eventually evolve from an air and space force, which we call ourselves today, into a **space and air force.**"[14] In a similar vein, a US Air Force historian, Richard P. Hallion, has also written, "We must dominate the military space dimension and integrate space force into our overall war-fighting capabilities across the spectrum."[15] Maj. Gen. Brian Arnold, director of Space and Nuclear Deterrence in the Air Force Secretary's Acquisitions Office, has stated "The air force strongly supports the Space Commission report and is already moving to implement many of (its) recommendations." Increased planning for future conflicts in space is evidence that these statements are not mere rhetoric. The air force has established a new Space Operations Directorate, started a Space Warfare School that consists of the 527th Space Aggressor Squadron and the 76th Space Control Squadron to develop and test space-based weapon systems.

In June 2001, the US Space Command and the US Strategic Command were merged to create a structure "responsible for both early warning of, and defence against, missile attacks as well as long range conventional attacks."[16] A Policy Coordinating Committee for Space has also been established within the National Security Council to provide an appropriate forum to develop, coordinate, and monitor space activities. More organisational changes are expected to augment military space capabilities, and work towards the enhancement of professional space military education at all levels to ensure a more direct integration of space activities in military operations.

Indeed, the present US Administration is preoccupied with turning the vision of space-based military operations into reality. It is engaged in the development of advanced space weaponry—in particular, the space-based laser (SBL) and space-based kinetic kill vehicles. The SBL is considered the "follow-on" technology to missile defence, and would deploy a constellation of 20-30 lasers orbiting the earth. Possibly powered by nuclear reactors, they would be tasked to destroy/disable other "competitor" satellites or targets on earth.[17] Most of these systems derive sustenance and support from a report entitled, "Joint Operations Superiority in the 21st Century," prepared in 1999 by the Defence Science Board. It had then identified certain advanced technologies that US military operations in 2010 and beyond would need to develop and induct into operations. These included two-stage ballistic-missile launched precision weapons for attacking high-value ground targets; GPS satellites used in conjunction with kinetic energy or conventional penetrator projectiles; a constellation of SBL to provide global coverage and defence against hostile missile launchers; and a fleet of space orbiting vehicles carrying rods of heavy material in highly elliptical orbits to reenter and transit the atmosphere striking targets at hypersonic speeds (Mach 10 or 10,000 feet per second).

The second major programme in this regard is working on improvements upon ASATs to arm them with a potential to destroy or disable a satellite. The USA and Russia have been developing ASATs since the late 1950s.[18] The first generation ASATs were either non-nuclear or nuclear ballistic missile launched weapons that were either direct hit-to-kill devices, or satellites that would explode in close proximity to the target satellite so that the resultant debris would destroy the target satellite. The new generation of ASATs includes Directed Energy Weapons (DEWs) that are space-based laser weapons using nuclear detonation in space to produce X-rays, neutrons, gamma rays, or other parts of the electro-magnetic spectrum that will then be directed at the target satellite.

It is, therefore, with a clear strategy in mind that the US budget today allocates $30 billion for a space weaponisation programme. In fact, there are reports that the Missile Defence Agency is attempting to have allocated a figure of $284.8 million for spending on SBL alone for the period 2003-2007. At the same time, it has also sought $1.33 billion for space-based kinetic kill vehicles over the same period.[19]

The continued development of space systems seriously undermines the claim that the US missile defence project is purely defensive in nature. Instead, the dual use capability of both the SBL and kinetic kill vehicles makes them inherently threatening to space assets and raises national security concerns of other countries. The further the United States pursues these systems, the greater pressure it places on international strategic stability, and hikes the possibility of sparking an arms race in space which would have seriously detrimental effects on global commerce, communications and security.

Implications of Space Weaponisation

Space weaponisation, along with BMD and use of precision weapons increases the temptation for preemption, particularly since the US has officially proclaimed preemptive strikes as part of its military strategy. These disquieting developments threaten to overwhelm the deterrent capabilities of adversaries anywhere on the planet. While this may be reassuring for the US domestic audience, it nevertheless raises the threat perceptions of countries across the world, ripples of which shall make themselves felt in many ways and regions in the coming years. International suspicion of the United States has been further heightened by its unwillingness to support efforts to establish a legal regime on PAROS. Despite overwhelming international backing for the regime, work on it has been stalled in large part due to the objections of the United States, which has argued strenuously that there is no need for a legally binding agreement. It has expressed its readiness to only discuss, not negotiate, an agreement banning weapons from outer space.

Among the countries that are carefully monitoring the US intransigence and the implications of BMD for weaponisation of outer space, China stands out. Realising the vulnerability of its relatively small posse of ICBMs to the NMD, China has consistently maintained that PAROS is the most urgent issue before the international community, even more than the negotiation of the FMCT. Chinese officials have used every opportunity and forum to press this issue and highlighted every aspect of the risks involved, from an arms race and conflict in space to the problems of space debris and crowding of space. It has expressed concern that space debris arising from the tests and deployment of space weapons would raise serious hazards for spacecrafts and satellites. Also, it holds that deployment of large numbers of SBW would crowd the LEO and limit the orbit resource usage for civilian purposes.[20]

Indeed, Beijing is acutely conscious of the fact that control of space and missile defence capability coupled with precision strike weapons would raise the possibility of US intervention in any conflict over Taiwan. It also finds these diversions counter-productive to its concentrated efforts in shoring up its economic development. In desperate attempts to counter this trend, at the diplomatic level China has repeatedly pointed out the lacuna in the existing legal regimes with respect to issues concerning the military use of space. In fact, in June 2002, China joined hands with Russia to circulate a joint working paper at the CD. Entitled "Possible Elements for a Future International Legal Agreement on the Prevention of the Deployment of Weapons in Outer Space, the Threat or Use of Force Against Outer Space Objects", the paper suggested banning SBWs and the threat or use of force against outer space objects.[21]

However, China is also fully aware of the limited impact of these diplomatic attempts on putting any brakes on the US space weaponisation programme. Consequently, it is reported to have accelerated its own programme for the exploitation of space for military purposes. China launched its first military communications satellite in January 2000 as part of the PLA command and control network linking forces for combat. These will be connected to airborne and ground-based sensors to give ballistic missiles, cruise missiles, aircraft and ships a seamless tactical-to-strategic targetting capability. Studies are under way to develop a Global Mobile Satellite Information System (GMSIS), which would provide personal hand-held communications via 18 to 24 satellites in medium orbits.[22] The PRC has been exploring the possibilities of micro-satellites, which are smaller than most satellites and, therefore, less expensive. A satellite in this category, Tsinghua 1, was launched on June 28, 2000, by a Russian booster.[23] Its launch has put China into the selected bracket of countries that can design and operate micro and nano-sized satellites. This success has implications for China's scientific programmes as well as its enhanced military satellite capabilities.

China has also begun testing a variety of anti-satellite measures. According to some published Chinese military "white papers", one kind of ASAT weapon that would be effective against a space-based laser weapon is being developed as part of China's asymmetrical warfare strategy "to fight and win a high-tech war against a powerful adversary", most notably the technologically-superior military capabilities of the United States.[24] According to a Pentagon Report, "China's PLA is building lasers to destroy satellites and already has beam weapons capable of damaging sensors on space-based reconnaissance and intelligence systems. Consequently, China could blind the US intelligence and military space equipment systems vital for deploying US military forces in current and future warfare."[25]

Russia too has taken every opportunity to decry weaponisation of space. It has argued in favour of the use of outer space in conformity with international law and in the interest of maintaining peace and security and promoting international cooperation. In this context, it has sought a multinational obligation not to put into the orbit around the earth any objects with any kind of weapons and not to install such weapons on celestial bodies or deploy them in outer space in any other way and not to resort to the use or threat of use of force with regard to space objects. It has also favoured the establishment of a mechanism to monitor implementation of this agreement, on the basis of confidence-building measures and transparency of space activities, including earlier submission of comprehensive information about space objects to be launched, inspection of sites from where space objects are to be launched, and creation of an appropriate international body to implement such procedures.[26]

American allies in Europe too are apprehensive of US domination of space and initial European reactions to NMD plans have been less than welcoming. Countries that are likely to host some components of missile defence have witnessed popular demonstrations against the projected deployments. For example, on March 4, 2000, some 300 protesters gathered at the Menwith Hill (UK) base, demonstrating against construction of the US-backed SBIRS system.[27] In May 2000, more than 270 British MPs had signed six separate Commons motions urging the government to have nothing to do with Washington's new anti-ballistic missile project. Greenland's Parliament declared in November 1999 that if US NMD plans violate the ABM Treaty, then Greenland "can't support plans for an upgrade of the Thule radar". Danish Foreign Minister Niels Helveg Peterson backed up this declaration on February 25, 2000, clarifying that a "firm component" of Denmark's policy is that use of the Thule radar not be "in violation of international rules."[28] These comments appear to reflect a serious gulf between US and NATO views on missile defence policy. Some European countries have also expressed the apprehension that the NMD and eventual space domination might decouple American security policy from Europe and it is significant that the European Space Agency is engaged in developing independent space capabilities for observation, remote sensing and navigation.[29]

Arms Control in Outer Space: Challenges and Options

Extending arms control into outer space in order to preempt and prevent attempts at its weaponisation poses a huge technical and political challenge. The relatively easy facility of crossing over from civilian use into military use of space, and the technical limitations in the actual implementation of space arms control complicate the scenario. The following paragraphs reveal the challenges in the path of arms control.

Dual Use of Space

The commercial use of space is a long-standing fact of life today. Satellite services today enable global communications, television broadcasts, weather forecasting, navigation of ships, planes, trucks, cars, synchronising computers, communications and electric power grids. According to a study by the Union of Concerned Scientists, the US has about 200 commercial and civil satellites in operation today, Russia has about 30, the European Space Agency has about 50 and Japan about 25 such satellites in orbit.[30] In fact, there are more commercial satellites in space than military ones.

However, there is a very close linkage between the commercial and military uses of satellites. For instance, satellites that are in use for commercial communications are also being used by the military for collecting information on capabilities and intentions of potential adversaries, navigation and weather forecasting. In fact, it is telling that the US Department of Defence has seen merit in using commercial satellites without investing too much of its defence budget towards military assets in space. Obviously, other countries too have done the same and, therefore, commercial satellites are being used equally for military purposes, including monitoring treaties and agreements.

This deep linkage between the civil and military uses of space now forecloses the possibility of designating space as a sanctuary in which all kinds of military activities, including reconnaissance, surveillance and communications, are banned. In fact, since the 1960s, military satellites have been tasked to perform communication functions, mapping, meteorological support, tracking enemy missiles tests, assisting in navigational support for each nation's naval forces, performing a vital strategic early warning function (warning of a possible nuclear attack by opposing forces), performing tactical warning and attack assessments, as well as spying on each other's militaries, leadership and strategic assets. Over the past few decades, advances in weapons technology have produced new, high-tech precision missiles and bombs, whose accuracy depends on the uninterrupted flow of information provided by satellites in space. These developments have further strengthened the dimension of space into military strategy and created the concept of military power in space. Just as military superiority on land, sea and air has become a critical element of current military strategy, supremacy in outer space has been projected as an essential element of successful warfare in the future.

As more and more countries realise the importance of synergy of space supremacy with land, sea and air superiority and aspire to full spectrum dominance through the development of capabilities to exploit this trend, the feasibility of arms control in space shall proportionately fall. In fact, even now the only arms control still possible in outer space is to prohibit the deployment of military strike vehicles for war in space. As argued by Coyle and Rhinelander, "Space weaponisation, if defined as dealing with strike vehicles but not sensors (such as SBIRS), is the classification that presumably would ban activities of greatest concern and where agreement might be possible at some point in time."[31] The need, therefore, is to restrain the use of space for offence or attack but not for knowledge.

Techno-Legal Constraints

It is difficult to separate military interests from commercial ones. A number of techno-legal limitations, therefore, present themselves for undertaking arms control in space. First of all, there are no accurate and precise figures publicly available for the number of commercial or military satellites that are in orbit at any given time. No definitive answers are available for the number of satellites that might be on standby or that have failed but are yet structurally intact and in orbit. The lack of this information would become an even greater source of misunderstanding and suspicion once space is weaponised, bringing with it the greater danger of space war.

At the same time, a number of other difficult technical and legal definitions and issues will have to be resolved for meaningful arms control in outer space. These include questions such as: Can the production, testing and deployment of anti-satellite weapons be distinguished from BMD? Is a ban feasible only on the use of ASAT weapons? How does one define the boundary between air space and outer space? What constitutes a space weapon or a component of a weapon? Is there a feasible or necessary distinction between low and high altitude ASAT? What constitutes testing of space weapons? Are sensors acceptable in space and only strike weapons need to be banned? These queries embody several technical and legal issues while also carrying the baggage of political compulsions. Finding answers to them, however, is critical to be able to envision arms control in outer space.

Another critical issue pertains to verification that would necessarily include availability of information, transparency and inspections. Pre-launch and on-site inspections of payloads would have to be a part of an international space regime that would also include registration to provide basic information on the specifications, details of the mission and proper licensing. The effective implementation of these measures, however, requires the willing cooperation of countries and would be extremely difficult to handle in the case of anything less than complete mutual trust and faith in the system.

Arms Control in Outer Space: Existing Treaties and Conventions

While the task of arms control in outer space may not be easy, solutions, however, can only be found when the international community accepts the value of doing so and its importance arises from three basic reasons:

- For maintaining international strategic stability and peace.
- For preventing an arms race or conflict in space. Once shooter or strike weapons have been deployed in space, the related problems of command, control, maintenance and retrieval would increase manifold, also raising the risk of their use due to miscalculation. An unintentional space war could well be triggered off by an on-orbit satellite fragmentation that may occur because of chemical change or pressure build up, but may be erroneously taken as an attack warranting retaliation.
- For continued and unhampered verification of arms control and disarmament treaties. Ironically, science and technology has given outer space an important role to play in disarmament and arms control. Space-based satellites and sensors are indispensable for monitoring and verifying major arms control agreements.

As of now, the Outer Space Treaty that entered into force in October 1967 provides the basic framework on international space law. It establishes the following principles:

- The exploration and use of outer space shall be carried out for the benefit and in the interests of all countries and shall be the province of all mankind.
- Outer space shall be free for exploration and use by all states.
- Outer space is not subject to national appropriation by claim of sovereignty, by means of use or occupation, or by any other means.
- States shall not place nuclear weapons or other WMD in orbit or on celestial bodies or station them in outer space in any other manner.
- The moon and other celestial bodies shall be used exclusively for peaceful purposes.
- Astronauts shall be regarded as envoys of mankind.
- States shall be responsible for national space activities whether carried out by governmental or non-governmental bodies.
- States shall be liable for damage caused by their space objects.
- States shall avoid harmful contamination of space and celestial bodies.

A further confidence building treaty, the Convention on Registration of Objects Launched into Outer Space (the Registration Convention), was negotiated in 1975. This established the practice of recording the launch of any space object into the earth's orbit in an appropriate national registry. The launching state was asked to furnish to the UN secretary general, the depositary of the Convention, information on the registration number, date and territory or location of the launch, basic orbital parameters and general function of the object. While this treaty complemented the 1972 Convention on International Liability for Damage Caused by Space Objects, neither agreement has received much attention or prompt observance. Rather, there have been inordinate delays in announcing launches and even where information has been provided, description of any military functions of the launched space object has been withheld.

In December 1979, in another initiative, the Agreement Governing the Activities of States on the Moon and Other Celestial Bodies (the Moon Agreement) was signed. It entered into force in 1984. It prohibits any threat or use of force on the moon or the use of the moon to commit any such threat or act against the earth, spacecraft or other man-made space objects. It also bans the placing of WMD on the moon and emphasises the freedom of scientific investigation on the moon. Meanwhile, the 1932 International Telecommunication Union (ITU) Convention, as amended in 1992 and 1994, protects civilian satellites from interference.

Evidently, these agreements regulate small sub-sections of outer space. There is no single treaty looking at these issues in a whole, composite fashion. Of course, space issues are also discussed in the United Nations by the Committee on the Peaceful Uses of Outer Space (COPUOS) attached to the General Assembly's Fourth Committee, and in UNISPACE. These forums provide a useful sounding board for concerns, but they have no power to regulate activities.

The CD in Geneva has long had an agenda item on PAROS, although the issue has failed to get into the CD's annual programme of work. PAROS issues have intermittently been discussed in the CD plenary and subsidiary bodies. But talks have often floundered on a basic conflict between the non-aligned states, who seek discussions and exploration of the issue to lead to appropriate agreements, and others—mainly Western states—who accept only exploratory discussions, perceive them as an end in themselves, and rule out negotiations for a new treaty. The most they have been ready to accept are confidence-building measures, instead of treaty negotiations, on the grounds that the end of the Cold War had brought about considerable changes and there was no longer an arms race in outer space.[32] For example, France had circulated a working paper in 1993 that called for confidence building measures in outer space to enhance the existing treaty regime, strengthen the security of space activities, prevent the aggressive use of space, and promote space cooperation for civil and scientific purposes.

In December 2000, UNGA Resolution 55/32 was introduced by Sri Lanka and co-sponsored by a number of non-aligned countries, besides Russia and China.[33] It reaffirmed the importance of the Outer Space Treaty, emphasised the need for consolidation and reinforcement, including verification, of measures to prevent an arms race in outer space, and called on countries with major space capabilities to refrain from acts contrary to maintaining a peaceful outer space. The resolution received 163 votes in favour and none against. The United States, Israel and Micronesia abstained. The same pattern continued in November 2001, when the UN resolution calling for space to be used for peaceful purposes summoned the support of 138 nations, but the US still cast a lonely abstention.[34]

At the CD in March 2002, China moved to strengthen the Outer Space Treaty, to "negotiate and conclude an international legal instrument banning any weapons, weapons systems and their components in outer space, with a view to preventing the weaponisation of outer space." China received wide support from other nations, but the US continued to maintain its position. Interestingly, in theory, no country, including the US, professes any opposition to PAROS. But when it comes to translating these political attitudes into a multilateral negotiation process leading to a treaty or convention, progress remains elusive.

Attempts shall have to be made to identify and tap constituencies that have a stake in banning weapons in space. In this regard, it may be useful to convince the civilian space industry to lobby for the prohibition of strike weapons in space. Since it has invested billions of dollars in commercial space assets for communication, broadcasting, weather forecasting or remote imagery, it would expectedly be against any disruption in their activities caused by a conflict or threat of a conflict in space. For instance, the US space-technology industries are estimated to have generated $125 billion in profits in 2000, and by 2010, the cumulative US investment in space is expected to reach as much as $600 billion, roughly the equivalent of total current US investment in Europe.[35] In the next decade, it is projected that about 300 military satellites, valued at over $35 billion, will be launched worldwide. In the same period, however, more than 1,000 commercial satellites, valued at over $50 billion, will also be launched.[36] Surely, the industry would not want to jeopardise investments of this sort by an arms race in space and would rather support its peaceful use as a universal objective. Therefore, the civilian space industry could be used to garner support against the weaponisation of space. The feasibility and acceptability of arms control in this realm, though, shall have to balance the commercial and military interests of nations.

Striking a Balance

It is no longer possible to declare outer space a sanctuary that disallows all actions of military utility. Reconnaissance, surveillance and communication for military purposes using the medium of space are now routine tasks performed by national space assets. Some contend that space operations are now a "pre-requisite for successful intervention on the ground, efficient communication and good coordination of precision strikes."[37] While the US is taking the lead in this, others with space capabilities can be safely expected to follow the trend in the coming years.

However, the line that needs to be drawn, and which may still be possible, is to ban the entry of strike weapons, whether conventional or those of mass destruction, and their delivery vehicles in space. It is not too late to prohibit weapons and regulate space activities to prevent offensive and aggressive deployments and activities. This would call for a fair amount of ingenuity, tactful diplomacy and innovative solutions. But approaching it with a positive attitude and the requisite political will can close the door on space weaponisation.

As a first step, it will be important to agree upon clear definitions and parameters on the types of activities that can be permitted in outer space and those that must be strictly regulated or prohibited. Since there are significant similarities between the technology and characteristics of forms of ASAT and BMD, the most basic question arises on the very acceptance or rejection of missile defence. While some wish to demilitarise space altogether by prohibiting even BMD, such a radical step no longer seems feasible given the present political conditions. The international community does appear to have accepted some level of missile defences. Therefore, at this stage, it would be more prudent to concentrate on the more feasible task of preventing future weaponisation by focussing on explicitly banning the use of ASATs, and on seeking agreement to regulate military activities.

Consequently, negotiations on a multinational agreement that could be called a Treaty to Prohibit Weapons and War in Space would have to comprise at least four main components:[38]

(i) A ban on the deployment and use of all kinds of weapons in space, thereby extending and strengthening the 1967 Outer Space Treaty's prohibitions on WMD in space. This would include laser and other directed energy weapons and kinetic energy weapons, as well as any other potential offensive innovations.

(ii) A ban on the testing, deployment and use of ASAT weapons, whether earth-based or space-based.

(iii) The establishment of a code of conduct that safeguards non-interference with clearly defined space assets and activities.

(iv) The establishment of a norm-based regime that provides incentives for compliance and severe penalties for violation.

Another important issue pertains to the forum that should conduct the exercise of negotiating such a treaty. While the CD might be the best equipped to do so, the stalemate on PAROS that has now existed for several years dampens the hopes for any optimism in this regard. Consequently, some analysts have suggested an independent route to start multilateral negotiations so that it removes PAROS from its blocking position in the CD and also ensures that prevention of weaponisation of space gets addressed sooner rather than later. Inevitably, the US can be expected to jettison such initiatives, but determined countries could consider the option of at least starting negotiations without US participation. If the initiative picks up momentum, then the US could be expected to join in the multilateral talks with a view to redirecting or manipulating them. It might also try to scuttle any agreements or treaties by refusing to sign. It is at this juncture that the determined political will of countries supporting the treaty would have to remain steadfast.

India and Outer Space

India could play a meaningful role in taking this process forward. In all its pronouncements on the subject, India has always maintained that outer space is the province of all mankind. Upholding the concept of common heritage of mankind, as applicable to the Antarctica, India, ever since the year 1982, has regularly tabled a resolution at the UN calling for a convention to be negotiated for prohibiting the use or the threat of use of nuclear weapons under any circumstances.

Realising that existing legal instruments are inadequate to deter imminent attempts for the further militarisation and weaponisation of outer space, India has emphasised the need for the CD to engage itself on this subject at the earliest. India, it may be recalled, was one of the co-sponsors of Resolution 55/32 that reiterated that the CD has the primary role in the negotiation of a multilateral agreement or agreements, as appropriate, on the prevention of an arms race in outer space in all its aspects.

The prospect of weaponisation of outer space that leads to the domination of this medium by the US cannot be in India's long-term interests. India has always upheld the need for multipolarity in international relations so that every country not only coexists in an environment of security, but also actually enjoys the sovereign right to make its own choices. This can be possible only when some countries do not have the power or capability to undertake unilaterally preemptive actions against what they describe as not in their national interest. India must continue its initiatives to prevent the unnecessary weaponisation of space that would not only be destabilising but also divert economic resources away from other more pressing tasks of development.

Conclusion

"...we shall not see space filled with weapons of mass destruction, but with instruments of knowledge and understanding..." So had said President John F. Kennedy in his Moon Speech on September 12, 1962. But that was then. Almost three decades later, the Rumsfeld Space Commission had concluded, "Space is not simply a place from which information is acquired and transmitted or through which objects pass...In the coming period, the US will conduct operations to, from, in and through space in support of its national interests, both on earth and in space."[39]

Indeed, perceptions and concepts have changed dramatically as technological capabilities have risen. So much so that today's US military policy appears to be more technology driven than based on any objective threat analysis. This is most evident in the US resolve to weaponise space, even though the threat to US satellites is not only a "low probability/ middle damage threat" but also one that could easily and more effectively be dealt with by universally banning the development of SBWs and ASATs. But the current American thinking and initiatives are moving in a direction that could hurtle the world down an extremely dangerous path.

The testing and deployment of space-based weapons would bring in its wake several far-reaching implications and would be militarily and politically self-defeating. Space weapons would not only be expensive, but also provocative and escalatory. As with the expected push on NMD deployment, fielding an ASAT weapon will only spur others to develop and deploy their own counter-measures. Any initial advantage gained by one country from being the first to put weapons in space would soon be neutralised as other major powers seek to develop space weapons of their own, while lesser powers could offset them with asymmetric responses. It must be recalled that the United States seriously underestimated the speed with which its advantage as the sole nuclear power in 1945 would erode. In less than five years after the US took the lead, the USSR and the UK had caught up with it. The same could be repeated with space weapons too. Several countries, including Russia and China, have developed sophisticated anti-satellite weapons, and several others are seeking such technologies. If these continue to proliferate, anti-satellite weapons would undermine fundamental US interests, and endanger international security and commerce.[40]

Testing and deployment of space-based weapons would complicate the use of space for other commercial and relatively "peaceful" military purposes. While the economically affluent and resource rich countries might be able to build and maintain such expensive assets, other less endowed countries too would be compelled to respond with similar moves. However, even a limited arms race of any kind would cast an enormous financial burden and, even more significantly, create a climate of mistrust and rivalry in international relations.

The weaponisation of outer space will also jeopardise the chances of the abolition of weapons of mass destruction. Some achievements in outer space technology are actively used for such "auxiliary" military purposes as communication, navigation, global positioning systems, etc. and help to maintain strategic stability in the world by ensuring compliance with disarmament arrangements. In future, the use of space-based means could help to resolve a broader range of issues of international security, such as, for example, nuclear and missile technologies proliferation control, conventional and strategic arms build-up in potential conflict areas, etc. But with weapons in space, trust levels between countries could plummet and more nations could see merit in acquiring nuclear weapons for offering at least some notional amount of deterrence.

At another level, the problem of space debris cannot be taken lightly either. Amidst the several hundred active satellites currently in orbit around the earth, there are also thousands of pieces of space debris. From spent rocket stages that never reentered the atmosphere to particles of propellant and corrosive byproducts of satellites, LEO has been described as a space "garbage heap." There are currently over 10,000 orbital objects larger than four inches (10 centimetres) in diameter that can be tracked by radar and optical telescopes. Of these, over 8,500 are catalogued by the US Space Command. Only five per cent are operational spacecraft. The rest is space debris. In addition, there are also tens or even hundreds of thousands of objects from 1 to 10 centimetres in size (half an inch to four inches in size) which are very difficult, if not impossible, to keep track of. The problem is that these objects travel at tens of thousands of miles per hour and represent a threat to active satellites and manned spacecraft.

Considering all of the above, it is imperative that the international community learns and avoids the mistakes of the Cold War. Nearly half a century ago, the nuclear arms race had taken on an irresistible momentum of its own and proved unstoppable. Precious resources had been squandered in the process and the world was perilously close to extinction. While this situation still persists, the weaponisation of space shall add another and even more dangerous dimension to it.

Therefore, there is an urgent need to develop a verifiable and enforceable international convention prohibiting the testing and placing in orbit of any object with a conventional, laser, or kinetic capability that could intentionally damage any other object in space or on the surface of the earth.

In 1985, the Sri Lankan Ambassador at the UN, Jayantha Dhanapala, had noted that preventing an arms race in outer space "is an easier task than attempting to control and decelerate such a race after it has begun." In 2003, there is an even greater urgency in this regard as the window of opportunity to keep space from becoming the site of the next arms race is fast closing. It is imperative now to translate the wider level of international commercial and military interests in the non-offensive uses of space into a transnational movement to prevent space from becoming a violent battleground of the future.

Notes

1. Interestingly, from a strictly legal point of view, there is no internationally acceptable definition of outer space. It is mostly understood to begin at an altitude of 100 miles, though in the US, astronaut wings are awarded even for flights that reach a height of 50 miles. Information as available from the US Space Command, "Down to Earth Space Questions", http://www.spacedaily.com April 9, 2001.
2. National technical means of verification are normally used to refer to photographic satellites in the low earth orbit that are used for information gathering on compliance with arms control measures.
3. The weaponisation of space essentially means the placement of weapons in orbit whose mission is to destroy/disable other satellites or other targets on land, sea or in the air. It would also include ASAT weapons.
4. PAROS has figured as an item on the agenda of the CD since 1982 as a confidence building measure that could promote transparency and predictability in the space activities of states.
5. LEO ranges from about 180 miles to 1,200 miles up.
6. Prepared Testimony on Ballistic Missile Defence to the Senate Armed Services Committee, Paul Wolfowitz, July 12, 2001 as available on http://www.defenselink.mil/speeches/2001
7. The Commission submitted its report to Congress on January 11, 2001. The text of the report is available on http://www.space.gov/commission/report.htm. Emphasis added.
8. Donald Rumsfeld, "Transforming the Military", *Foreign Affairs,* May-June 2002.
9. Statement made by John Bolton, US under-secretary of state for arms control and non-proliferation at the CD and as quoted by David Grahame, "A Question of Intent: Missile Defence and the Weaponisation of Space," BASIC note, May 1, 2002, available on http://www.basicint.org

10. US Space Command website document as quoted in Martin Malin, "Arms Control in Search of Constituencies," unpublished paper presented at the 52nd Pugwash Conference on Science and World Affairs, La Jolla, California, August 10-14, 2002.
11. Edy Korthals Altes and Alyn Ware, "Weaponisation of Space," May 4, 2001.
12. As reported in the *International Herald Tribune,* of January 30, 2001, and at the BASIC website.
13. As quoted by Grahame, n. 9.
14. Jeffrey Mason, "Space: Battleground or Frontier of the 21st Century," Centre for Defence Information, http://www.cdi.org. Emphasis added.
15. Ibid.
16. Malin, n. 10.
17. For more on this see, *Militarisation of Outer Space,* Document no. 7 prepared by the International Peace Bureau (Geneva) and available at website http://www.ipb.org
18. The first successful satellite intercept reported as a close intercept occurred on May 24, 1962, and was launched by the US from Kwajalein Atoll.
19. Grahame, n. 9.
20. Malin, n. 10.
21. The paper can be found at http://www.basicint.org
22. As quoted in A.V. Lele, "China as a Space Power", *Strategic Analysis,* 2002.
23. It was a joint project of Tsinghua University of Beijing and Surrey Satellite Technology Limited (UK).
24. As reported in the *Worldnetdaily* of January 20, 2001.
25. Lele, n. 22.
26. Georgy E. Mamedov, deputy minister of foreign affairs of the Russian Federation, at a conference on "Space Without Weapons" in Moscow on April 11, 2001.
27. See London Press Association, "UK Anti-Nuclear Protesters Target US Military Base," March 4, 2000, *Foreign Broadcast Information Service (FBIS-WEU-2000-0304).*
28. Gordon R. Mitchell, "US National Missile Defence: Technical Challenges, Political Pitfalls and Disarmament Opportunities," *ISIS Briefing Paper no. 23* (London), May 2000.
29. Austria, Belgium, Denmark, France, Germany, Ireland, Italy, Netherlands, Norway, Sweden, Switzerland, Spain and the UK are among the members of the European Space Agency.

30. Philip E. Coyle and John B. Rhinelander, "The Path to Controlling Weapons in Space", unpublished paper presented at the 52nd Pugwash meeting 275 on Science, Sustainability and Security at California, August 10-14, 2002.

31. Ibid.

32. Rebecca Johnson, "How to Move Forward: NGO Approaches and Initiatives for Addressing Space Security," in James Clay Moltz, ed., *Future Security in Space: Commercial, Military, and Arms Control Trade-Offs*, Monterrey Institute for Non-Proliferation Studies and Mountbatten Centre for International Studies, Occasional Paper No. 10, July 2002.

33. Traditionally, the lead in PAROS in the CD and UN Resolutions has been taken by Sri Lanka, together with Egypt. In 1985, as pressure increased from US plans for SDI, Sri Lanka's Ambassador to the CD, Jayantha Dhanapala, had noted, "We have watched [PAROS] in its transition from what appeared to be a slice of science fiction to being perhaps the single most important development in the arms race today". That was sixteen years ago, and his words are even more true today.

34. Karl Grossman, "Arms Race in Outer Space," article syndicated to Progressive Media Project and available on http://www.pma.org

35. Grahame, n. 9.

36. Mason, n. 14.

37. Therese Delpech as quoted in Gotz Neuneck, "Prospects for Space Arms Control", unpublished paper presented at the 52nd Pugwash meeting on Science, Sustainability and Security, at California, August 10-14, 2002.

38. For a detailed view on this, see Rebecca Johnson, "Multilateral Approaches to Preventing the Weaponisation of Space," *Disarmament Diplomacy*, Issue no. 56, April 2001.

39. Report of the Commission to Assess United States National Security Space Management and Organisation, January 11, 2001, n.7.

40. The commercial implication of any disturbance in space can be gauged from the failure in May 1998 of a single Galaxy IV satellite that caused 80 per cent of the pagers in America to go dead, affecting some 37 million users. Information as cited in an article by Michael Krepon in *Foreign Affairs*, 2002.

8

Reducing Nuclear Weapons, Increasing Nuclear Dangers

Manpreet Sethi

The year 2002 was greeted by the United States with the unveiling of the conclusions of the NPR.[1] Amongst other highlights, the proposal to bring about a substantial reduction in the US nuclear arsenal stood out. Upholding President Bush's announcement of November 2001, the NPR projected a scaling down of operationally deployed warheads from the present figure of nearly 7,000 to 3,800 by 2007, and their further reduction to a figure between 1,700-2,200 by 2012.[2] Therefore, over the next ten years, the US nuclear arsenal is expected to become several times smaller than the huge pile that it was at the peak of the Cold War.[3]

Normally, one would have expected such an announcement to be greeted with enthusiasm and optimism, raising the hopes that more countries would reciprocate in kind. At least Russia should have jumped at the offer, considering that it has been suggesting reductions in nuclear forces to 1,500 since its economic conditions do not allow it to maintain the present high levels. Not much, however, was forthcoming from anywhere on the US announcement. President Putin, too, in a lukewarm response, only mentioned that Russia would consider if it could respond in kind. China was restrained in its reaction. Neither were the proponents of nuclear disarmament ecstatic despite the fact that every plan of universal nuclear disarmament envisages a systematic reduction of nuclear weapons as a crucial step. The lack of enthusiasm for the latest US nuclear reductions may be attributed to the developments that have been projected as likely to accompany this measure. These threaten to nullify its benefits, and may prove to be immensely unsettling for international security.

What are these developments that threaten to lessen the value of nuclear reductions? This chapter explores these in order to illustrate the worthlessness of reductions if they come with the kind of baggage that they appear to be bringing along. In such case, from the point of view of universal nuclear disarmament and better international security, the reductions mean little. Rather, they are likely to adversely impact on the future of global nuclear dynamics, strategic stability and inter-state relations. The chapter concludes by highlighting that as a means to the larger goal of an NWFW, reductions in the numbers of warheads would be useful when they bring about simultaneous reductions in the threats, in the demands for WMD, and in the sources of insecurity.

Developments Accompanying the Cutbacks

The US announcement on nuclear reductions is a travesty of sorts. While it cannot be denied that by itself, every reduction is a welcome step, the matrix in which the current cutbacks are offered have reduced the initiative to little more than a rationalisation of the US nuclear arsenal based on a contemporary review of its interests and security requirements. Envisaged as part of a larger package that includes several other measures that would increase nuclear dangers, the proposed nuclear reductions may well prove to be worthless from the point of view of devaluing nuclear weapons in national military strategies, and rather gravely provocative to greater proliferation. The changes likely to come about in the composition and character of the US nuclear force, coupled with a deployed missile defence, and potential military control of space have serious implications for the shape and future direction of the international political environment, and the wider issue of nuclear deterrence. The result of all this is more likely to push other countries towards refurbishing and expanding their own nuclear forces in order to restore a semblance of strategic stability. The following paragraphs summarise some of these developments that the US NPR recommends along with the scaling down of the size of the nuclear arsenal.

Constitution of a Responsive Force

While the NPR proposes an unprecedented cutback in the US nuclear warheads, it stops short of recommending a complete destruction of the dismantled weapons. Rather, it suggests that the downgraded weapons be moved to a new category of the US nuclear force structure, a Responsive Force, where they would remain intact in reserve, and be ready for recall if the president so desires.

This practically implies little more than a reduced operational availability of nuclear warheads and not their complete non-availability. Critics have dubbed this as "deceptive counting" that makes nuclear reduction a "mere mirage".[4] The weapons shall augment an already existing "hedge" of some 2,500 active and inactive nuclear weapons.[5] The warheads in the hedge serve three purposes: they act as replacements for warheads destroyed each year in routine reliability and safety tests; they provide a safeguard against failure of operationally deployed weapons; and, a majority are intended to maintain the capability to increase the size of the operational arsenal quickly by "reconstituting" or "uploading" retired warheads onto nuclear missiles and bombers.

The hedge has grown substantially since the implementation of START I, and the proposed reductions shall add to it. But the US sees this force as a precaution against a dangerous and uncertain future. Post-September 11, this fear of insecurity in the US has only grown as is reflected in the premises on which the NPR was built:[6]

- A prevailing atmosphere of strategic change.
- The existence of multiple potential opponents, sources of conflict and unprecedented challenges.
- Varying and unequal risks and stakes as arising from 12 nations with a nuclear weapons capability, 28 armed with ballistic missiles, 13 with biological weapons and 16 with chemical weapons.

In such a situation, the US justifies the need for scaling down its operational warheads provided they are placed in reserve as a responsive force that would be ordinarily inaccessible, but still available, if required.

However, it must be pointed out that weapons in reserve are mostly unaccounted. This increases the opacity of the nuclear arsenal. Lack of transparency could then engender mistrust, making reciprocal reductions more difficult and less probable. In fact, this trend of shrinking deployed and accountable warheads while adding them to an unaccounted reserve force could perpetuate a dangerous transformation of nuclear stockpiles by pushing them into clandestine postures.[7] In the absence of a clear picture about the levels of one another's nuclear forces and aggregate warhead inventories, verifiability shall be an even more sensitive and complicated issue in any future arms control.

Built-in Reversibility

Since the NPR categorically places nuclear weapons as a hedge and envisages their recall when necessary, there is an in-built reversibility in the unilateral nuclear reductions that decreases their usefulness for promoting greater confidence among other countries. In the absence of the permanent destruction of weapons, others are likely to interpret the initiative as little more than a move to temporarily store extra warheads in another location.

Of course, several within the US establishment have justified the need for a flexible nuclear arsenal for two reasons. One, so that the force is able to quickly respond to the changing strategic, technical, operational and political environment, thereby providing greater adaptability to the president. And, two, because the US Congress is more likely to accept such an initiative, rather than one that deprives the president of a "broader portfolio of response and options."[8] The US Congress does prohibit the president from reducing the country's nuclear warheads more than the cutbacks stipulated in the START Treaties. Notwithstanding these US justifications, the bottom line remains that the fear of reversibility of the US initiative is unlikely to encourage other countries to follow suit.

Lack of Transparency and Verification

Reductions that are not codified in a treaty reinforce the non-binding status of the cutbacks and the possibility of their reversal. This assumes even greater significance in view of the fact that over the last half a decade, the US Administration has shown an aversion to being tied down by treaties that hinder its freedom.

While unilateral reductions help avoid long, detailed and complicated negotiations on complex verification regimes that consume immense time and effort, nevertheless, it must be reiterated that arms control agreements are useful tools for generating inter-state confidence and, hence, are worthy of the time and effort spent on them. A legally binding document that underpins implementation of the reductions, particularly monitoring and verification, provides assurances and safeguards as regards the others' nuclear forces.

In this context, several analysts have suggested that it may be more useful if the unilateral US reductions can eventually be connected to a mutually acceptable cooperative regime. This could be done by either extending the provision of START I verification and monitoring to a unilateral reduction process in what may be called "cooperative unilateralism",[9] or through some other sort of arrangement between the countries. Responding to this criticism, the US had negotiated the Strategic Offensive Reductions Treaty (SORT) with Russia in May 2002, but as is discussed in the subsequent section, the treaty does little more than legally bind the two countries to the reduced number of warheads. On several other issues, it has fallen far short of expectations.

Likely Development of Mini-Nukes

In his official briefing on the NPR, US Assistant Secretary of Defence J.D. Crouch pointed out that there are no recommendations in the report about developing new nuclear weapons. But more significant is his qualification of this statement with, "We are trying to look at a number of initiatives," including modifying existing weapons to give them "greater capability against...hard targets and deeply buried targets."[10]

The reduction in numbers, therefore, is likely to be accompanied by a simultaneous effort to produce new weapons, especially small, high precision and bunker-buster types. As of now, a 1994 legislation prohibits laboratories from undertaking research and development of such weapons, but Press reports have suggested that the issue would figure in the coming sessions of Congress. It is well known that weapon laboratories, as well as some politicians and military officials have long pushed for low yield weapons that could be more usable, especially to defeat hardened and deeply buried targets such as bunkers housing terrorists, or biological and chemical weapons facilities. The US military experience with Iraq over the last decade, and its more recent battle against the Taliban who used caves as their hideouts can only strengthen the case for these mini-nukes. A report published by the National Institute for Public Policy in January 2001 also supports such weapons for their "unique targeting capabilities," especially against chemical and biological weapons in states of concern. The same case has been made by the director of the Sandia Laboratories, "...the yields of weapons left over from the Cold War are too high for addressing the deterrence requirements of a multipolar, widely proliferated world. Without rectifying that situation, we would end up being self-deterred."[11]

Supporters of mini-nukes further argue that "precise" nuclear strikes with low yield warheads would minimise, if not obviate radioactive fallout and, hence, restrict collateral damage. Nothing could be farther from the truth. Even a precise low yield nuclear weapon cannot be expected to carry out as surgical a strike as is being proclaimed. A report prepared for the Federation of American Scientists (FAS) claims:

> No earth-burrowing missile can penetrate deep enough into the earth to contain an explosion with a nuclear yield even as small as one percent of the 15 KT Hiroshima weapon. The explosion simply blows out a massive crater of radioactive dirt, which rains down on the local region with an especially intense and deadly fallout.[12]

The mistaken perception, therefore, that mini-nukes cause less collateral damage could prompt their use and thereby break the norm against nuclear first use. The deployment of battlefield nuclear weapons would perceptibly lower the threshold for nuclear attacks. But a break in the taboo against the use of nuclear weapons and the grant of legitimacy to nuclear war would be disastrous for international security.

Augmenting Test Site Readiness

On the issue of nuclear testing, the NPR reiterates opposition to the American ratification of the CTBT, and, at the same time, recommends that the country should reduce the time necessary to prepare its nuclear test sites from the current period of two years. This is despite the fact that the US has an SSMP that is a virtual nuclear testing facility. The programme ensures the continued viability of the US nuclear weapon infrastructure, trains a new generation of weapon scientists, and takes up refinement of weapon designs and production processes. In fact, weapon scientists have suggested that the programme has the capability to meet emerging military requirements through new weapons that "could be deployed…without the need for nuclear tests."[13]

Yet, the NPR proposes accelerating test site readiness. This, however, is a move fraught with danger. For one, doing so would require an "additional expenditure of $15 million annually for three years for the Nevada test site."[14] More importantly, though, it shall strain the test moratorium in place worldwide, dim hopes or the eventual entry into force of the CTBT, and push other countries to factor nuclear tests in developing or modifying their nuclear weapons by placing a question mark over US intentions in the matter. It may be recalled that in the year 2001, President

Bush had even suggested that the US would understand if the Chinese were to undertake further nuclear tests in the face of an NMD deployment. The conduct of nuclear tests by any country, however, could open a Pandora's box. Others might be pressured to follow suit in order to restore their perception of strategic stability.

NMD Deployment

President Bush's plans to build a missile defence system now that the US has withdrawn from the ABM Treaty shall certainly undermine the significance of nuclear reductions for disarmament. While the weapon count may reduce in the American arsenal, it is likely to increase in other countries as they reassess their vulnerabilities and respond by adding more warheads to their nuclear forces.

Dismissive of such ramifications, however, the NPR outlines missile defence as a crucial third leg of the New Triad, the other two being the development of a combination of non-nuclear strike capabilities to enhance deterrence; and, the storage of extra weapons as a hedge against uncertainties. As explained by Crouch, these moves are aimed at creating a "broader array of capabilities...to deter and, if deterrence were to fail, to defend against potential adversaries."[15] This viewpoint derives from President Bush's speech of January 2001 wherein he had stated, "Deterrence can no longer be based solely on the threat of nuclear retaliation," and, hence, the need for "new concepts of deterrence that rely on both offensive and defensive forces."[16]

Measures such as these, however, adversely impact on international security. Russia, China, Europe, India and several other countries have expressed concern at the potential impact of the NMD on global security. The Russian government has hinted at maintaining the option of arming its SS-27s with MIRVs, as well as maintaining existing MIRVed ICBMs outlawed under START II. In any case, this treaty has now been pronounced dead and its replacement in the form of SORT is silent on the issue of MIRVs. At the same time, Russia is likely to accelerate the development of counter-measures such as decoy-shrouded warheads, and anti-satellite weapons to defeat missile defences.

China too has used every international forum to criticise the move and urged the US to give it up. But, in the face of US intransigence, Sha Zukang, director of the Chinese Foreign Ministry's Department of Arms

Control and Disarmament has warned "If not, we'll be ready."[17] Indeed, China has lost no time in factoring the US NMD into its strategic and security calculations. Extremely conscious of the adverse impact that a reliable missile defence would have on its small strategic deterrent missile force, and, hence, on Taiwan,[18] China's strategy in this regard encompasses three main elements—enhancement of its nuclear deterrence capabilities through military modernisation; alignment with like-minded countries; and, increase in its bargaining power by holding out threats of proliferation of WMD technologies (and also counter-measure technologies in the future) to countries perceived as states of concern by the US.

China's military is engaged in an ambitious programme of nuclear and missile modernisation. Research and development is on to graduate to a new generation of solid-fuelled, multiple warhead and road-mobile missiles that would reduce the launch preparation period.[19] It is also estimated that China may succeed over the next 10 to 15 years in deploying a viable "second leg" of its deterrent in the form of SLBMs. A report prepared by the National Intelligence Estimate of the US Central Intelligence Agency (CIA) forecasts a significant nuclear weapons build-up by China between 2002-2015.[20] The next decade is also likely to see further improvements in China's command, control, reconnaissance, and early-warning capability, including the possible introduction of space-based assets to support these functions and vastly increase the accuracy of its missiles. In another kind of response to the US NMD deployment, China has also threatened to indulge in nuclear and missile proliferation to other countries. The adverse implications of this for world security cannot be underestimated.

Despite triggering off a chain of adverse reactions, the NMD, however, may still not be able to guarantee absolute security to the US. A certain amount of residual uncertainty would always remain. Meanwhile, ballistic missiles are also the least likely means by which terrorists would deliver a weapon against the United States on account of the difficulties involved in developing, acquiring, or using an ICBM. Rogue states too would be foolish to use missiles because they provide a known point of origin, which would likely result in immediate US retaliation.

Therefore, while the guarantee of security that the NMD shall provide is less than absolute, the impact on inter-state relations is certain to be not all benign. Suspicious attitudes and actions premised on mistrust can naturally be expected to breed more of the same. In addressing its threat perceptions, the US may well end up upsetting existing international security arrangements and fostering a worldwide lack of confidence.

Weaponisation Of Outer Space

Along with the Rumsfeld Commission report that assessed the missile threats to the US, and on which the NMD is premised, in the late 1990s, the US Congress had also commissioned Donald Rumsfeld to prepare another report on outer space and its relevance to US security. This report urges the US to "develop weapon systems that operate in orbit and augment air, land and sea forces."[21]

Indeed, in its drive to build an unparalleled defence force, the establishment of US dominance over outer space occupies a preeminent position. Some of this is envisaged through an eventual upgradation of the NMD that would anyway incorporate satellite-based sensors to detect and track missile launches and targets, and guide the interceptors. But the US is also funding more such development. A statement of Keith Hall, assistant secretary of the air force (space) before the US Senate Committee on Armed Services Subcommittee on Strategic Forces, on March 8, 2000, revealed:

> ...[I]n preparation for the transition to an Aerospace Force, the Air Force S&T (science and technology) community has more than doubled its S&T investment in "space-only" technologies from about 13 per cent in FY 1999 to 32 per cent a year by FY 2005. This increase in investment is primarily focused in five areas: space-based radar, space-based lasers, reusable launch vehicles like the space operations vehicles, satellite survivability, adaptive optics, and hyperspectral imaging...[22]

US efforts in this direction have caused maximum concern to China that has elaborately spelt out the dangers of militarisation of outer space at every forum. In fact, negotiations for the conclusion of the FMCT remain stalled owing to China's insistence that an ad hoc committee be instituted to simultaneously consider the issue of PAROS, a move that the US is in no mood to concede.

While at the moment, no country may find itself with the financial and technological wherewithal to match American ambitions in the area, US moves to dominate space, nevertheless, do nothing for international confidence building. Rather, they shall vitiate international security, deal a blow to the philosophy of arms control, and lay the foundation for a potential arms race.

Ambiguity Between Strategic and Tactical Nuclear Weapons

The proposed nuclear reductions are planned for the operationally deployed strategic warheads, but do not seek to touch the 1,600 odd strong tactical nuclear force. This category of weapons was not covered by any of the two START Treaties either. However, their continuance in the arsenal does pose risks of accidental use or a miscalculated launch. Moreover, as tactical nuclear weapons are easier to transport, they are at a greater risk of theft or accident. This fear has been raised more in the case of Russia where thousands of tactical nuclear weapons are believed to be not subject to stringent control or foolproof accounting procedures.

In order to reduce the risks related to this problem, it would have been useful if the US had led the way in reducing these weapons as well and encouraged Russia to follow suit.

Predominance of US Unilateralism

An NMD that would enhance US invulnerability, backed by its domination of outer space, a formidable conventional and nuclear arsenal, a policy of counter-proliferation, and a disregard for multilateral treaties are indications of the American tendency to arbitrarily and unilaterally decide its national interests and define its code of conduct. Iraq and Yugoslavia already exist as two instances where the US chose to act without UN authorisation. More recently, in the case of Afghanistan as well, UN support was acquired after the US had declared its clear intention of launching military strikes against the Taliban government for its support to the Al-Qaida terrorists. Subsequently, the global war against terrorism has allowed the US greater command of the world, and fundamentally reshaped the architecture of international alliance.

In order to further shore up its capabilities, the US has announced a 15 per cent increase in its military spending. This shall be the biggest jump in the military budget in 20 years, and more than double the military spending in all of the European Union. There is a danger that such overwhelming emphasis on military power will further entrench Washington's unilateralism and lead to a retreat from the attempt to build multilateral institutions for international conduct. Naturally, this is viewed as a disturbing trend and could set into motion a chain of proliferation as each country tries to secure itself in any which way it deems fit.

SORTing It Out

By way of meeting some of the criticism on its unilateral approach to nuclear reductions, the US had agreed by February 2002 that it would codify the cutbacks through a bilateral agreement with Russia. On May 24 of the same year, during a summit meeting between Presidents Bush and Putin, the SORT was concluded, an arms control treaty arrived at after nearly a decade. The primary objective of the treaty is to legally bind the two countries to a reduction of their nuclear arsenals to a range between 1,700-2,200 by 2012. The basic significance of the treaty is derived from the fact that it binds the two countries to a level of reductions that go beyond those envisaged by START II to levels contemplated for START III. But, apart from this, the treaty has little on offer.

Rather, the treaty might be interpreted as a depiction of the political reality of the time in which Russia and the US are making a pretence at arms control. The new treaty provides a kind of tool to the US to showcase its readiness to bind itself to an arms control process that it was being condemned for having jettisoned when it announced its withdrawal from the ABM Treaty. SORT helps to deflect this criticism and will especially come in handy during the NPT PrepCom meetings when the US shall have to prove its commitment towards Article VI of the NPT.

Meanwhile for Moscow, the treaty helps to preserve the impression that the US still considers Russia its equal and wishes to negotiate a level of nuclear parity. At a time of deteriorating economic and conventional military strength, the inability to hold the US to its commitments under the ABM Treaty was a blow. Of course, Russia, as also the rest of the world, had reconciled to the inevitability of the deployment of the NMD, but by pinning the US to SORT, Russia has at least been somewhat comforted. Not coincidentally, SORT was concluded just a few days prior to the termination of the six-month period that would have freed the US from the ABM Treaty. With SORT ratified, the Russians can lower the numbers of their warheads that would have come down for economic reasons in any case, with the consolation that the Americans too are doing the same.

Apart from these generic and political benefits of the treaty, SORT makes little headway towards genuine arms control. In fact, the US has managed to keep the treaty brief (unlike START II that it seeks to replace, SORT is all of only 500 words long) and ambiguous so as to meet any future eventualities. While this ensures flexibility for the US nuclear posture, international security, however, could prove to be a casualty since the treaty may not be able to promote confidence among other countries. In terms of what the treaty embodies, and what it does not, it does not inspire faith in the likelihood of a nuclear weapons free future.

First, while stipulating the range of numbers of warheads that the US and Russia would arrive at by 2012, SORT allows each side to determine the manner, sequence and timetable that it would follow. In the light of this ambiguity, it has been calculated that for the immediate future, the number of nuclear weapons in the US arsenal might actually increase from the levels to which they would have dipped under START II.[23] But, ignoring this reality, the treaty lays down no scheduled reduction targets other than establishing 2012 as the deadline, which ironically enough shall also be the year of the expiry of the treaty unless the two countries mutually agree to extend it. This strange provision could reduce the treaty to no more than an eyewash providing a semblance of arms reductions, if the two parties decide to abandon the exercise a decade later.

Second, the irreversibility of the reductions provided by the treaty is in serious doubt for two reasons. One, it provides no stipulation as regards the manner in which the dismantled warheads should be disposed of. It is left to the discretion of the country as to what warheads it decides to eliminate, and whether the warheads removed from deployment are actually destroyed or not. The flexibility in this regard has obviously been maintained to uphold the US desire to reduce deployed weapons, but stockpile them in a reserve or responsive force. At the second level, the new treaty contains no provisions for limiting delivery vehicles. Rather, it allows the two countries to retain as many delivery vehicles as they like—without exceeding the START I limit, which itself will remain in force until 2009 only, and expire before SORT reaches its target. This flexibility along with the ambiguity on the storage of warheads makes reduction under SORT easily reversible in the future. Although neither START I nor START II mandated warhead destruction, both treaties required the dismantlement of delivery systems to make it harder for either superpower to redeploy warheads removed from service quickly. The START III framework that President Bill Clinton and Boris Yeltsin had agreed to in 1997 even visualised measures to destroy warheads to make reductions difficult to reverse. SORT, however, has sidestepped both these provisions, raising scepticism over the irreversibility of the process.

Third, SORT does not include the elimination of MIRVed missiles, a fact that makes it take a step back from START II that had stipulated the prohibition of this category of missiles in what had been hailed as a major achievement. This oversight is obviously a concession to Russia since it deploys a greater proportion of its strategic warheads on MIRVed ICBMs. Under the START II provisions it would have had to eliminate the practice of arming land-based missiles with MIRVs. In order to maintain strategic stability with the US then, Russia would have had to build a number of expensive new single warhead ICBMs. Under SORT, however, it can continue to maintain its earlier stockpiles and also MIRV the new Topol M (SS 27) missiles that it has recently added to its arsenal. Meanwhile, the absence of a legal prohibition on MIRV capabilities is sure to provide an impetus to the development and improvement of similar technologies in China. The cascading adverse effect of this move for the Southern Asian region cannot be overlooked.

Fourth, SORT has circumvented the issue of the removal of tactical nuclear weapons, a subject that was to have been addressed by START III. While it has been categorically stated that the START process is being abandoned, neither country has committed itself to the eventual dismantling and destruction of tactical nuclear weapons. The significance of this has already been discussed in an earlier part of this chapter.

Fifth, the US and Russia have not laid down any verification procedures to establish the veracity of the cutbacks. The onerous and complicated task instead has been left to a Bilateral Implementation Commission that is expected to meet at least twice a year to deal with issues of compliance. Displaying an unprecedented level of trust, the two have dismissed any need for transparency any more than this. But as has already been stated, verification and mutually agreed upon enforcement procedures perform an important task of confidence building.

In another move towards ambiguity, and in a clear deference to the American position, the new treaty regards the nuclear posture of each country as its own business. This, nevertheless, is a dangerous move because the NPR of any country that lays down the role of its nuclear weapons, its force posture and level of readiness, etc., has a distinct bearing on other countries. In this sense, it cannot be the business of only one state but has implications for regional and international security that cannot be ignored.

Seventh, SORT contains no specific counting provisions. The text of the treaty refers to only "operationally deployed warheads" as worthy of being counted. According to the Bush Administration, this includes those warheads assigned to ICBMs, bombers, and submarines that are in active service, but not to warheads associated with delivery vehicles that are being overhauled or undergoing repairs. Also, the United States holds that it can deploy delivery vehicles that can carry several warheads, but only those warheads actually mated to the delivery vehicle at any given time will count toward the treaty's limit. In other words, if a 10-warhead ICBM has only one warhead actually on it, it would count as only one warhead even if nine warheads stored nearby could be loaded onto the missile relatively quickly. Obviously, the American interpretation of the counting rules excludes warheads removed from active service and put into storage and/or warheads on delivery vehicles undergoing repair or not ready for launch. This provision, therefore, will allow many more weapons to remain at each country's disposal as long as they are on dealerted status.

The Russian standpoint on the subject, however, is slightly different, though it was the American view that eventually prevailed. While agreeing that the limit refers to deployed (as opposed to stored or reserve) warheads, Russia had sought rules along the lines of START I that counts warheads according to the maximum number deployable on a delivery vehicle. Had this been accepted, a deployed missile that could carry 10 warheads would have been counted as 10 warheads regardless of how many warheads were actually on it.

While the treaty does not specifically address the counting issue, effectively allowing the United States to hold its more liberal interpretation, the Russian Foreign Ministry has indicated that the issue would be discussed further in the meetings of the implementation commission. The outcome, however, may weigh heavily in favour of the US position.

On the issue of withdrawal from the treaty, each party "in exercising its national sovereignty" may withdraw from the treaty with just three months written notice to the other. The withdrawal clause in SORT greatly differs from that in START that required the party to withdraw only after explaining how "extraordinary events" had "jeopardised its supreme interests" to necessitate a withdrawal.

Table 7.1. **Comparative Assessment of SORT, START II and START I**

Parameter	SORT	START II	START I
Date of Signature	May 24, 2002	January 3, 1993	July 31, 1991
Date of entry into force	After ratification in May 2003	Treaty has been pronounced dead	December 5, 1994
Number of warheads permissible	1,700-2,200	3,000-3,500	6,000
Stance on delivery vehicles	None	• Deactivation by 2003 • No MIRVing • ICBMS and their launchers to be taken off alert	• Reduce to 1,600 by 2001 • Destroy all ground-launched IRBMs. • Ban on modernisation and replacement of heavy ICBMs and SLBMs
Counting rules	Ambiguous, applicable only to operationally deployed weapons.	Specific provisions for counting delivery vehicles and warheads	Specific provisions for counting delivery vehicles and warheads
Verification provisions	None except for two meetings a year of the Bilateral Implementation Commission	Use of national technical means, on-site inspections and monitoring activities	Use of national technical means, on-site inspections and monitoring activities
Withdrawal clause	Three months notice	Six months notice	Six months notice

On the whole, then SORT has managed to allow the US to unshackle its nuclear planning and posture from the linkage that existed between START II's entry into force and the constraints on missile defences. While it had already indicated its resolve to abrogate the ABM Treaty, it offered SORT as a fig leaf to Russia and the international community to soften the blow by codifying arms cuts in a treaty. The reductions, meanwhile, had already been conceived by the Bush Administration as a rationalisation of its arsenal in the face of new threats. SORT has not been able to sort out any of the other dangerous and destabilising moves that are likely to accompany the reductions. Instead, and as has also been interpreted by the Russians, the treaty has given a free hand to both sides to play their nuclear cards as they want, so long as the numbers remain in the stipulated range. In the ultimate analysis, however, the US might have made a bad bargain, trading enhanced security for flexibility.

Implications of the Accompanying Developments

Nuclear Proliferation

Notwithstanding nuclear reductions, the fact that they shall lead to an improvement in the US' offensive and defensive capabilities is certain to trigger further nuclear proliferation. An overtly proclaimed continued US dependence on nuclear weapons in clear violation of its promise under Article VI of the NPT, and a flexible arms control commitment through SORT from behind which US unilateralism still shows, are moves that are likely to spur countries that foresee a difference of interests with the US to secure their own national interests through the acquisition of nuclear weapons.

Traditional wisdom, as well as empirical observations of the Cold War era suggest that the only effective way of deterring WMD threats is by the acquisition of a similar capability by the adversary. The balance of terror then establishes a sort of stability and peace, however uneasy and precarious its foundations may be. Therefore, US insistence on retaining nuclear weapons while also erecting a missile defence are moves that shall reinforce faith in deterrence, especially for the smaller states. This, then, could signify nuclear and missile proliferation.

In fact, the prospects of proliferation are heightened by the increasingly easy accessibility to nuclear material, technology, expertise or even the weapon itself owing to the wide-scale dismantlement of several categories of nuclear weapons in both the US and Russia. As more warheads are dismantled but not destroyed under SORT, the possibility of their falling into the hands of "rogue" nations or sub-national actors cannot be dismissed. It must be considered that a Russia that is struggling to maintain a nuclear force can ill afford to devote scarce funds to properly storing retired warheads. These warheads could consequently become a tempting target for groups like the Al Qaida. From a proliferation standpoint, therefore, the treaty compounds the post-Soviet danger of "loose nukes". SORT will add weapons to Russia's vast complex of nuclear warhead and weapons-usable-material storage sites, some of which are reported to lack sufficient physical security, follow inadequate accounting procedures, and are protected by underpaid guards who could succumb to temptation. Additional warheads in these storage sites will increase demands on a security system that is already financially and technically challenged and could be a potential source of nuclear weapons for terrorists or rogue states. Given the reality of international terrorism today, the nuclear threat to the US from these undeployed but undestroyed weapons is far greater than to any other nation.

If the threat of nuclear terrorism is increased by the decision to hold onto dismantled weapons, the threat of nuclear proliferation through the transfer of weapons technology could arise from the policies that China decides to follow to even out US unilateralism and missile defence. China has revealed its intention to be "uncooperative in all multi- and bilateral arms control fora if the US proceeds with NMD".[24] In any case, Beijing is acknowledged to have a long history of selling nuclear and ballistic missile technology to other countries. But, with particular reference to the NMD, China may seek to restore the strategic balance by being even more liberal in providing nuclear and missile technologies to those who so desire. In fact, apart from purely strategic reasons, commercial considerations may also dictate such a policy as China tries to recover the some of the costs of building counter-measures. In order to recover some of these costs, China could see economic (besides strategic) sense in selling the relatively older technologies to other countries. Nuclear proliferation, therefore, could be on the upswing.

Faith in Multilateral Treaties

In order to facilitate the lawful deployment of the NMD, the US has withdrawn from the ABM Treaty. This sets a precedent of sorts and impacts on the fate of international treaties, especially those relating to non-proliferation and arms control. For instance, if a treaty like the ABM that is widely believed to be the cornerstone of all arms control agreements can be scrapped by the US, then it impinges upon the sanctity of such international agreements because other countries may be tempted to do the same when it no longer suits their interest to abide by them. Until now there is no example of a country walking out of its commitments agreed to under a treaty. But others could pick up the US examples of the treatment meted out to the ABM Treaty, the CTBT, the Biological Weapons Protocol, the NPT, the Kyoto Protocol, etc.

Nuclear Disarmament

From 1997 onwards, US official thinking has revealed a growing propensity for an overt reiteration of the centrality of nuclear weapons in US military strategy. The 1999 NATO nuclear doctrine upheld the same and Russia responded likewise. China too looks upon nuclear deterrence as the necessary assurance for its security. And India, and subsequently Pakistan also, have come to accept and establish nuclear deterrence. Further proliferation would lead to a more widespread acceptance of the doctrine and thereby deal a severe blow to the hopes of realising a world without nuclear weapons.

The US NPR had provided an opportunity to put an end to this reliance on deterrence through a substantive devaluation of nuclear weapons. However, as is now evident, the US has chosen to only reduce numbers while adding value to those that remain. In the bargain, nuclear disarmament may have been pushed further off the agenda as other nuclear weapon states busy themselves in crafting adequate responses to the emerging threats.

Conclusion

Considered by themselves, the US and Russian nuclear reductions constitute a constructive move. Unfortunately, though, they come as part of a larger package that takes away the positive sheen. By raising insecurities, leaner but infinitely meaner future nuclear arsenals threaten to trigger an increase, vertical or horizontal, in the count of nuclear weapons elsewhere. This appears even more likely since these words of the US president appear to echo the sentiment in other NWS as well: "Nuclear weapons still have a vital role to play in our security..."[25] An overall increase in the valorisation of nuclear weapons is unlikely to reduce their attraction for countries. How and why should it be expected that countries would be ready to acquiesce to the existence of these WMD in the hands of few and also suffer silently as they erect a defence shield without wanting to secure themselves as well? Questions of national security are only going to rise in the face of a blatant display of unilateralism by the US in its interactions with countries it perceives as threats, and in its treatment of multilateral treaties that it feels are getting in the way of its national security. National and international security analysts shall certainly closely study the manner in which the case of Iraq is handled in the coming years.

These are not trends that engender confidence and could be counter-productive for global security, as others feel compelled to invest undue resources on activities aimed at securing themselves. Further horizontal nuclear proliferation surely cannot be in the interest of any nation. But the means to arrest it are to be found not in export controls or missile defences, but in a world order that fosters mutual security and in which countries have faith in multilateral mechanisms of interaction—whether political, economic or military (conventional/nuclear). In order to build such a world order, it may be prudent to take steps towards increasing the "transparency and irreversibility" of nuclear arms reductions by subjecting non-deployed arsenals to international scrutiny and mandating their destruction so that a rapid reconstitution of nuclear forces would not be possible. In fact, the value of reductions would be further enhanced if they were to be complemented by the following:

- Reduction in fissile material.
- Reduction in threats.
- Reduction in demands for WMD.

Nuclear reductions are essentially a means for nuclear disarmament and that is how they must be envisaged. In the long run, a gradual move to zero through the involvement of others NWS may prove to be the best form of defence. By reinforcing non-proliferation, this would uproot the threat right at the source—an objective infinitely cheaper, safer and more effective than an untested missile shield or an uncontrollable proliferation spiral.

Notes

1. The NPR is a comprehensive review of US strategic nuclear policies, doctrine, force structure, command and control, operations, supporting infrastructure, safety, security, and arms control. The Congressionally mandated report also examines the selection of targets, stockpile levels, and new and potential threats to the United States and its allies and interests worldwide.
2. Figures as presented by the US Department of State in a report dated July 2001.
3. Maintenance of the present stockpile costs the US exchequer $27 billion a

year, a figure that does not include the cost of upkeep, stability and security of the arsenal. Current proposals to reduce the weapons could save between $920 million to $8.3 billion over the next ten years. Figures as projected by the Congressional Budget Office and quoted in Press Release by Taxpayers for Common Sense, April 30, 2001.

4. BASIC, "NPR Point-Counterpoint," http://www.basicint.org
5. The present hedge comprises mainly retired warheads from the 1987 Intermediate-Range Nuclear Forces (INF) Treaty and the 1991 START I accord, which required destruction of delivery vehicles (bombers and missiles) but not the warheads.
6. As derived from the official briefing of the US Department of Defence, dated January 9, 2002, and made available by BASIC on http://www.basicint.org
7. It is estimated that before START I, about 5 per cent of the total stockpile was in the inactive category. Plans for the START II stockpile could increase that ratio to a 1:1 ratio, with the reserve constituting as large a stockpile as the deployed stockpile.
8. Assistant Secretary of Defence J.D. Crouch as quoted in BASIC, n. 6. He was also explicit in stating that at all stages of the proposed reductions, the US would "retain the option to slow, stop or reverse the process." Also see Mark Bromley, "Pentagon Touts Nuclear Cuts While Escalating Risks," Global Beat Syndicate, http://www.nyu.edu/globalbeat/syndicate/
9. The term has been used by Rose Gottemoeller in an article entitled "Offense, Defence, and Unilateralism in Strategic Arms Control", *Arms Control Today,* September 2001.
10. BASIC, n. 6.
11. As quoted in Robert W. Nelson, "Low-Yield Earth-Penetrating Nuclear Weapons," *FAS Public Interest Report,* January-February 2001, http://www.fas.org.
12. Ibid.
13. Director, Sandia Labs as quoted in "US Nuclear Weapons Policies, BMD, and the Quest for Weapons in Space: Military R&D and the New Arms Race," *Western States Legal Foundation Information Bulletin,* Summer 2000.
14. John Harvey, director of Office of Policy Planning Assessment Analysis at the National Nuclear Security Administration, calculated this figure as quoted, "US Testing: White House Says Nuclear Testing in Future Option," *Global Security Newswire,* January 11, 2002.
15. n. 6.
16. President Bush's speech on missile defence, delivered at the National Defence University, Washington, on May 1, 2001, and as available on website http://www.acronym.org.uk
17. For more on this, see Manpreet Sethi, "Ballistic Missile Defences: Implications for India," *Strategic Analysis,* vol. XXV, no. 6, September 2001, pp. 737-752.

18. China is most vehemently opposed to the provision of systems such as the PAC-3 or Aegis-equipped naval vessels, that might overtly link US and Taiwanese defence capabilities in what China would view as a revival of the pre-1979 Washington-Taipei mutual defence treaty.
19. The first of these new missiles, the Dong Feng 31, was tested in July 1999 and more efforts are on to improve the survivability of these missiles.
20. *A New York Times* report cited by Charles D. Ferguson and John E. Pike, "Will Missile Defence Shoot Down Arms Control?" *Yaderny Kontrol (Nuclear Digest),* vol. 5, no. 4, Fall 2000.
21. K. Subrahmanyam, "National Missile Defence," *Diplomatic Corps,* vol. V, no. 2, May-June 2001.
22. As quoted in *Western States Legal Foundation Information Bulletin,* n. 13.
23. Even as extended in 1997, that treaty would have taken the two sides down to 3,500 warheads by 2007, and relevant delivery vehicles would have been deactivated by 2003. Under the new treaty's timetable and the NPR, the United States will still have some 3,800 warheads in 2008. John Holum, "Assessing the New US-Russian Pact," *Arms Control Today,* June 2002.
24. Ambassador Hu Xiaodi's statement at the CD as quoted by Sverre Lodgaard, "European Views of the US NMD Programme," *Pugwash Occasional Papers,* vol. 2, no. 2, March 2001, p. 50.
25. n. 16.

9

Devaluing Nuclear Weapons: Difficult, But Desirable

Manpreet Sethi

The Swing of the Pendulum

During the long decades of the Cold War, strategic analysts of both superpowers justified the continuance of nuclear weapons for the maintenance of strategic stability between the two blocs that were ideologically so different from one another. Jealously guarding their respective spheres of influence, and guarantors of the security of other countries in their respective camps, the US and the USSR indulged in competitive and often hostile political interactions and could hardly conceive of the possibility of abandoning the concept of nuclear deterrence. Even though nuclear weapons were never brought into actual use, incidents abound where their use was subtly threatened. Nevertheless, as and when it suited their interests, the superpowers did negotiate certain arms reductions treaties to reduce their nuclear arsenals to more rational levels. Not for a moment, though, was there any serious consideration of the prospect of their abolition since the WMD were perceived to be of immense value.

The dictates of the Cold War sustained the nuclear weapons between 1945 and 1990. But once this period came to an end in the early 1990s, much hope was expressed that nuclear weapons would lose their importance. This optimism hinged on the fast developing cooperative relations between the US and Russia[1] after the collapse of the Soviet Union. So high were these expectations that a few international strategists almost began writing obituaries of war as they forecast a declining utility of nuclear weapons.[2] Interestingly, experts from a diverse range of backgrounds—former military leaders, diplomats, scientists, strategists, and academics—published convincingly optimistic cases in favour of nuclear disarmament and recommended that US nuclear policy seriously consider the task. One notable report, published by the Washington based Stimson Centre, recommended, "The ultimate objective of US national security policy should be the elimination of all weapons of mass destruction from all states—not the preservation of nuclear deterrence in perpetuity."[3] At a more international level, the Report of the Canberra Commission on the Elimination of Nuclear Weapons laid down a step-by-step plan for an NWFW.

These prognostications, however, had underestimated how deep the roots of nuclear deterrence had penetrated during the Cold War years. Today, more than a decade after the fall of the Berlin Wall, and despite such strongly worded reports and statements in favour of nuclear disarmament, all the NWS continue to rely on these weapons as a cornerstone of their national security for the "indefinite future". Rather, the US, UK, NATO and Russia have since updated their nuclear doctrines to assign newer roles and missions for their nuclear arsenals. For each one of them, these words of Under-Secretary Walter Slocombe, though spoken in the context of the US in his testimony before the International Security, Proliferation and Federal Services Sub-Committee of the Senate Government Affairs Committee on February 12, 1997, ring equally true: "Nuclear deterrence, far from being made wholly obsolete, remains an essential, ultimate assurance against the gravest of threats."[4]

If earlier the Cold War had justified their existence, it is now the end of the Cold War and the emergence of new threats that appears to vindicate the continued existence of nuclear weapons. In fact, once again efforts are on to tailor present day threats to defend the existence of nuclear weapons. Therefore, if post-September 11, terrorism has been identified as the main threat to international security, the US NPR released in 2002, among other recommendations, makes a case for the development of bunker busting, mini-nuclear weapons that can penetrate deep into the earth to decimate underground terrorist hideouts. Undoubtedly, the lessons of fighting a battle in the caves of Afghanistan have been learnt quickly enough. Similarly, another contemporary threat identified in errant missiles being launched from rogue states is sought to be met through the NMD. But the consequent impact of this development on the rest of the countries by way of accelerating a nuclear arms race is ignored. The implications of the BMD have already been examined in an earlier chapter, but it would not be amiss to reiterate that most countries perceive the NMD as a US design to seek strategic superiority with scant regard for the cost in terms of a disruption of global and regional strategic balance and stability. In order to restore some sort of balance, these countries are likely to increase their dependence on nuclear weapons and seek their qualitative or quantitative improvements.

Developments such as these add value to nuclear weapons and raise the chances of proliferation. It is a sad reality of our times that the stock of nuclear weapons is on the ascendant, belying all the hopes of the early 1990s. Arms control and nuclear disarmament are beginning to fade from the agendas of countries that matter in this regard, as developments described earlier in a chapter on nuclear reductions steer the US policies. The renewed emphasis on nuclear deterrence, however, is certain to impose a heavy cost on international security.

With a scenario such as this looming large on the international horizon, how does one identify steps for devaluing and delegitimising nuclear weapons? At a juncture when the value of nuclear weapons is apparently on an upward swing and it is the stock of arms control and disarmament that is fast depreciating, how does one arrest the trend? This chapter attempts to answer these questions. It suggests certain steps to devalue and delegitimise nuclear weapons that could serve as stepping-stones to eventually arrive at universal nuclear disarmament. The complexity of the problem and the fact that it is intertwined with issues of state sovereignty, power play, etc., make it impossible to expect that all countries would suddenly, in one leap, arrive at a comprehensive treaty on nuclear abolition. Conditions to make this possible will have to be gradually worked out and created. The devaluation strategy being put forth hopes to strip nuclear weapons of their utility and render them redundant and unusable through a series of measures so that the ground would have been prepared for their eventual elimination.

A caveat, though, is in order since the strategy wades through a palpable sense of *déjà vu*—of it all having been said before, to little avail. However, a conscious effort is made to keep the feeling of cynicism in check and the realisation up front that the inability to act urgently and decisively would inevitably result in dangerous and far-reaching repercussions for international security and possibly for the future of mankind.

Devaluation is Desirable

Before identifying steps to actually devalue nuclear weapons, one needs to be convinced that devaluation will actually be a smart move. This is imperative because the desirability of undertaking this exercise has time and again been questioned by those who vouch for the benefits of nuclear deterrence in keeping a third world war at bay.[5] The proponents of this school of thought see these weapons as security enhancers since they threaten such horrendous consequences that the adversary is deterred against making any move that could invite their use. According to this logic, nuclear deterrence provides stability to the international order. This belief has over the years contributed to the inertia against realising nuclear disarmament.

When subjected to closer scrutiny, however, the belief reveals itself as no more than an untested and untestable hypothesis. Of course, nuclear weapons have deterred the use of nuclear weapons. But their presence has done nothing to really stop conventional conflagrations. These did take place and the presence of nuclear weapons seemed not to matter in most of these cases. For instance, in Korea or Vietnam or Afghanistan or the Gulf War nuclear weapons were not used, and neither was their presence able to deter the very act of aggression. In the case of Vietnam, it is especially worth noting that the US chose to lose a humiliating and destructive war against a small, non-nuclear country even as its nuclear stockpile sat ready. In more recent times, the experience of India and Pakistan during the Kargil War of 1999 is also illustrative of this since it involved two newly armed states that are geographically contiguous and whose relations are characterised by constant border skirmishes. Nuclear weapons failed to deter the Pakistani infiltration or the Indian retaliation, and escalation to aerial bombardment. This experience overturned the hypotheses that nuclear powers do not go to war against one another.

In fact, if it is conceded that nuclear weapons reduce the chances of large-scale conventional conflicts between nuclear-armed powers, then, would it not be logical to encourage nuclear proliferation as a means to obviate major conventional wars? But it would be sheer fallacy, if not downright suicidal, to indulge in such a line of thinking since it is a fact that wars have been, and will continue to be, a means of settlement of disputes among nations. Man killed man in the Stone Age and he will continue to do so into the e-tech future. But it is the presence of nuclear weapons that threatens to escalate a confrontation to the nuclear level. This then is a frightening prospect because a nuclear war would cause destruction of life on a massive scale and with relative ease.[6] In fact, the destruction so wrought would be measurable not only in terms of the number of casualties, but would go much further in time and space.[7] Given this reality, it is indeed ironical that we seem to be depending upon our potential agents of extinction as guarantors of our security. An NWFW will definitely be a *safer* world and to that extent, it is more *desirable*.

Risks from Nuclear Weapons

Nuclear War

The very presence of nuclear weapons poses three major risks that cannot be taken lightly. First, there is the danger of a conflict involving one or more NWS escalating to a nuclear level. Besides this horrendous possibility, the presence of nuclear weapons also carries the threat of a nuclear war being triggered off by an unintended accident or miscalculation. In fact, modern day communication and sophisticated delivery systems take pride in extremely short lead times between the political decision to use nuclear weapons and their ability to reach the pre-designated target. This carries the inherent risk of a nuclear weapon being launched during a crisis without adequate thought, or in the heat of the moment. The danger of unauthorised nuclear launch cannot be dismissed either, since however advanced the nuclear command and control structure of a country, it cannot be absolutely foolproof.

Nuclear Proliferation

The second risk that nuclear weapons carry is the seed of their proliferation. If those who possess the weapons continue to zealously guard their arsenals, then over time, others too could be expected to feel the need to develop the same. It is not going to be possible to indefinitely profess that a nuclear arsenal is vital to the national security of a few but harmful to global security if others develop or acquire the same. It is in this context that Jonathan Schell has referred to the US arsenal as a "proliferant" and not as a "deterrent,"[8] because as long as weapons exist with one, others are going to aspire to possess them.

The prospects of nuclear proliferation are further eased in contemporary times given the widespread availability of information on how to construct a nuclear bomb and the relatively easy accessibility to fissile material. Of course, the NPT or the export control regimes provide some sort of a notional barrier. But, as the examples of Iraq, North Korea and Pakistan[9] have demonstrated, neither the NPT, nor other export control regimes, are able to restrain those not ready to voluntarily accept limitations on themselves.

While one cannot deny any country the right to defend itself, even if it aspires for nuclear weapons if it so deems necessary, nuclear proliferation does impose additional strain on international security and stability. Nuclear weapons, by their inherent nature are not weapons of just any other kind. Their possession has to be necessarily accompanied by the development of a fairly sophisticated command and control apparatus so as to minimise, if not totally obviate, the chances of accidental, miscalculated or unauthorised nuclear war. And, it is for this reason that non-proliferation of nuclear weapons becomes a desired objective. But non-proliferation can be sustained only when it is accompanied by a devaluation of nuclear weapons and their eventual elimination from security considerations.

Nuclear Terrorism

A third risk of nuclear weapons is manifest in nuclear smuggling or terrorism. This too has been brought into sharper focus by two recent developments. First, the incidents of September 11, 2001, have drawn attention to the very real threat of international terrorism fuelled by religious fundamentalism as a tool to wage proxy wars. A well-established nexus among international terrorism, organised crime and narco-trafficking ensures a steady supply of financial resources, the latest weaponry and motivated men. Highly inspired and monetarily well-endowed outfits then, might have little compunction about using a nuclear weapon, or its threat, for certain limited gains.

The second factor that raises the probability of nuclear terrorism is the less than adequate storage facilities for fissile material and weapons at the Russian nuclear weapon complexes. With the breakdown of centralised Soviet control, instances of smuggling of small amounts of radioactive material have been periodically reported. The problem is further compounded by the state of economic deprivation that the unemployed or underemployed workers at these storage sites face. This makes them easy prey for temptation from a sufficiently motivated and cash-rich terrorist outfit. While there may be little logic in nuclear terrorism, nevertheless, ease of accessibility could prompt its use, especially since such outfits are not likely to always act within the confines of rationality.

Given the dangers that persist with the existence of nuclear weapons—deliberate nuclear wars; miscalculated, accidental or unauthorised nuclear exchanges due to uncontrolled proliferation; and nuclear terrorism—the desirability of devaluing them with the ultimate objective of doing away with them should now be self-evident.

Stepping Onto the Ladder

Steps intending to reach a state of universal nuclear disarmament have often got lost in a maze of questions that have no easy answers. Problems of logistics, issues of state security and sovereignty, the matter of political will are only a few of the complicating factors. The matrix is made even more complex by the basic chicken and egg dilemma: should the security concerns of countries be addressed first through the creation of an international system that is based on peaceful coexistence and the rule of law enforced through world governance? Could the demand for these weapons be quelled only after a complete resolution of all differences between states and in a world where all nations live in complete harmony? Several strategists have argued that disarmament would require inter-state relations to be robust and trustworthy. In fact, political relations would have to be so good as to ensure the non-existence of any adversarial relationship.[10]

Undoubtedly, a world that has renounced armed conflict as a means of settling disputes and in which countries live in perfect harmony would indeed offer the ultimate security guarantee and remove all justification for such weapons. The challenge, nevertheless, lies in conceiving nuclear abolition in this real world inhabited as it is by "sovereign states with conflicting ideologies."[11] The task, of course, cannot be expected to be easy by any standards. At one level, it calls for exemplary political vision and statesmanship that has the courage to discard entrenched beliefs and accept new norms of inter-state behaviour that are more transparent and hinged on more participative institutionalised mechanisms. At another level, it requires a resolve on the part of the military to shake off the crutches of nuclear deterrence and accept seemingly lower levels of security, especially by states that are conventionally weaker and see nuclear weapons as a leveller of sorts. Logistically, it calls for large doses of innovative expertise to deal with the discarded weapons.

It is keeping the magnitude of these problems in mind that certain steps are suggested in the following paragraphs to devalue nuclear weapons. These are not necessarily sequential and, in fact, more than one can be pursued simultaneously. Each would, nevertheless, complement the other, contribute to confidence building and create an environment making it easier to take another.

Altering the Belief Systems

It has been said, rather wisely, "Since wars begin in the minds of men, it is in the minds of men that defences of peace must be constructed".[12] Indeed, the power of the human mind cannot be underestimated. In the ultimate analysis, it is beliefs and value systems that make man act the way he does. We are what we believe. It has been stated several times over that technology itself is neutral. It is the purpose to which it is put that makes all the differences. And the purpose is decided by a thinking man whose values and beliefs determine his behaviour and actions.

Applying this logic to nuclear weapons, beliefs that deify nuclear deterrence for its war prevention capabilities provide a rationale to these WMD. Since the majority of those who matter in strategic decision-making have believed in the undesirability and impracticality of nuclear disarmament, the belief has become entrenched. In fact, its roots have struck so deep that significant changes in the international situation in the wake of the collapse of the Soviet Union have still not brought about changes in thinking on these issues. Instead, new threats are being projected to justify the same nuclear philosophy that perpetuates the relevance of WMD.

The task, in fact, is further complicated by the existence of an extremely powerful military-industrial complex in every nuclear weapon state that has developed a stake in maintaining the status quo on nuclear weapons. Their very existence depends on obtaining huge amounts of public money to maintain and update nuclear arsenals. Consequently, and in a task that shall not be easy, political leaders in favour of nuclear abolition will have to stand up to the combined pressures from the military, the industry and other vested interests that wish to keep alive the existing beliefs and attitudes.

Therefore, a change in mindset and belief systems would be an important step towards devaluing nuclear weapons. Only when mindsets are strategically attacked to alter beliefs that ascribe value to nuclear weapons, can it be expected that nuclear weapons would loosen their stranglehold on national security doctrines. In order to break the mould, the belief systems need to be replaced by those more favourable to nuclear abolition.

History provides ample examples of institutions whose abolition was once thought to be unthinkable but that broke down under the pressure of changed belief systems. The case of slavery, for instance, is suitably illustrative. In fact, when the demand for abolition of slavery was first articulated, there were several who argued against it on the grounds that the economic and social systems would not be able to survive the drastic change!! At the time, on December 1, 1862, President Abraham Lincoln, in his famous message to the US Congress, had said, "The dogmas of the quiet past are inadequate to the stormy present...As our case is new, so we must think anew and act anew. We must disenthrall ourselves."[13] This advice has proved useful whenever radical transformations have taken place. And, it is most relevant today in the case of nuclear weapons. We have to disenthrall ourselves from their hold and allow new beliefs to replace the old. This would then create the necessary climate in which other more tangible actions can become possible.

Rewriting Nuclear Doctrines

National security doctrines are normally conditioned by factors such as the international strategic environment impinging on a country's security, an assessment of internal threats, its economic strengths, the military equipment and expertise available within the country, and its belief systems. As one or more of these parameters change, so do national security concepts. In the wake of the widespread changes brought about in the 1990s, most countries did undertake the exercise of reexamining and redrafting their nuclear doctrines in the light of changed threat perceptions. However, the absence of a simultaneous change in belief systems made all of them conclude that nuclear deterrence was still required to remain a principal pillar of their national security.

The annual report of the US Department of Defence released in March 1996, after the US had managed to bulldoze NPT member states into granting indefinite and unconditional continuity to the NPT, categorically stated that notwithstanding the transformed international scenario in the post-Cold War period, "strategic nuclear deterrence remains a key US military priority." Therefore, for the US, nuclear deterrence, far from becoming obsolete, is still upheld as an "essential, ultimate assurance against the gravest of threats."[14] The Russians, likewise, expressed, "For Russia, at least for the next several decades, there will be no alternative to nuclear deterrence."[15] In fact, on February 15, 2000, Sergei Ivanov, the secretary of the Russian Security Council, in a special briefing for the heads of the diplomatic missions in Moscow stated, "Russia considers the use of nuclear weapons to be a political means of deterring aggression...Russia *commits everything to the goal of deterrence* and will use the whole power of the state to administer a resolute and firm rebuff to the aggressor."[16] In 1999, NATO too in its newly articulated Strategic Concept rather staidly emphasised the role of nuclear weapons as "make(ing) a unique contribution in rendering the risks of aggression against the Alliance incalculable and unacceptable. Thus, they remain essential to preserve peace."[17]

Statements such as these provide a positive feedback to nuclear deterrence and reinforce its hold over strategic thinking. The strategy of deterrence then becomes self-perpetuating. It is this feedback loop that shall have to be broken in order to devalue nuclear weapons and for this a reassessment of the nuclear doctrines becomes imperative. The importance of this step lies in the fact that the role of the nuclear weapon is defined by a country's nuclear doctrine. Therefore, a rewriting of nuclear doctrines must be encouraged to make nuclear weapons less attractive by restricting, and eventually obviating their role in international politics.

Restricting the Role of Nuclear Weapons

Nuclear weapons were originally conceived and developed to deter the use or threat of use of other nuclear weapons. Gradually, however, countries detected in them the power to offset even conventional inferiority. Throughout the Cold War years, NATO banked on its nuclear arsenal to deter an overwhelming Soviet conventional might. In a strange twist of events, it is Russia, in the post-Cold War period, that has increased its reliance on this weapon in view of its substantially reduced conventional strength. Its national security doctrine put forth in 2000 affirms Russia's "right to use all available means and forces, including nuclear weapons, in case of the need to repel an armed aggression when all other means of settling the crisis situation have been exhausted or proved ineffective."[18]

For Russia, therefore, its nuclear arsenal performs two crucial roles at this present juncture. First, it fills an internal void and gives Moscow the requisite strength to pursue domestic policies of the kind that it deems fit. This is amply evident in the manner in which references to Russia's nuclear strength have been made while responding to international criticism of its policy in Chechenya. Second, the Russian nuclear weapons fortify an otherwise weak nation against the relative strength of other international players that Moscow perceives as security concerns.

In performing both these roles, the Russian nuclear weapons send out some messages: one, that these weapons can be effectively used to match and deter superior conventional capability. It is this belief that has led to the weapon being considered as one "of the weak against the strong, as the only weapon that can counter the conventional superiority of the West".[19] Second, these weapons grant prestige and status, as also the right to independent decision-making. This viewpoint is best expressed in the words of the French Prime Minister Alain Juppe, who had acknowledged in 1995, "By acquiring nuclear capability, France was able to play, on the world scene, a role well above that justified by its mere quantitative significance."[20] These perceptions, then, enhance the utility of nuclear weapons and motivate others to strive for them.

To a large extent, it was to remove the attraction of nuclear weapons as a means of assuaging insecurity arising from conventional inferiority or to counter the perception that they brought great power status, that the NWS had extended negative security assurances (NSA) to the non-nuclear states party to the NPT. Under UN Security Council Resolution No. 984 of 1995, the NWS reiterated their commitment not to use or threaten to use nuclear weapons against non-nuclear countries. However, a present US trend in favour of the expansion of the role of nuclear weapons to cover threats from other WMD such as chemical and biological weapons (CBW) undercuts the credibility of the NSA. Documents of the US Joint Chiefs of Staff clearly contemplate nuclear retaliation against the use of CBW. Therefore, the current US doctrine is premised on the freedom to *first use* of nuclear weapons to cover a number of contingencies, including "to deter use of non-nuclear weapons of mass destruction."[21] Such a formulation then could compel countries to reassess their own need for such a weapons capability to act as a leveller of sorts.

While the effectiveness of nuclear weapons to deter the use of biological weapons (BW) remains limited, because the source of a BW attack may be difficult to ascertain, making retaliation difficult, the expansion of the role of nuclear weapons may actually encourage the retention or acquisition of BW by countries that perceive these as an effective and inexpensive counter to the nuclear weapons of others. Besides, if a country as powerful as the United States insists that it needs nuclear weapons to counter non-nuclear WMD threats, then many less powerful countries, especially in conflict prone regions will be tempted to conclude that they have an even greater need for nuclear weapons for this purpose. A policy of nuclear deterrence of CBW would provide incentives and an easy justification for nuclear proliferation. Many other countries face far more plausible and immediate CBW threats than the United States. If US policy points to nuclear weapons as the ultimate answer to CBW, other states could have an increased motivation to acquire nuclear arsenals. Indeed, the worst outcome of a nuclear response is the prospect that it might be seen as militarily successful, thus, inspiring renewed belief that the perceived efficacy of nuclear weapons warrants their retention or, worse, acquisition.

In order to discourage this trend, the emphasis should be on developing other means to deal with the CBW challenge such as universalising the Chemical and Biological Weapons Conventions, which outlaw both classes of weapons and reinforce a taboo against their use. International pressure—United Nations resolutions or sanctions and other means—can also be brought to bear on states claimed to be producing, or about to use, such weapons. Under no circumstances should the nuclear weapon be conceded a multi-role capability. Rather, the attempt should be to devalue the weapon by narrowing its role to the core mission of nuclear deterrence and nothing beyond that. Once this happens, the motivation to acquire these WMD would substantially recede.

Strengthening the Nuclear Taboo

It is widely acknowledged among nuclear policy analysts as well as military practitioners that a nuclear taboo exists at the global level. This is based on a widespread popular revulsion against nuclear weapons and widely held inhibitions on their use. The normative presumptions against nuclear use, in fact, derive some legitimacy from an array of international arrangements such as nuclear weapon free zones, arms control agreements, negative security assurances and general laws of armed conflict. Unfortunately, though, there is no explicit legal prohibition outlawing nuclear weapons use and until such an instrument is created, it is important to sustain the norm of non-use.

In fact, strengthening the normative non-use of nuclear weapons would be a most useful step towards their devaluation. Norms circumscribe the realms of possibility of influencing behaviour, thus, determining the probability or non-probability of the occurrence of certain courses of action. The nuclear taboo, by delegitimising this class of weapon, decreases the likelihood of its use. It might even be true to state that in several cases it was this norm, and not deterrence alone, that stopped the use of nuclear weapons since they were considered unacceptable and immoral. This pattern has persisted over half a century and if it were to sustain over a longer period, it would reinforce a pattern of disuse that would be difficult to break. In fact, as the taboo gains strength, nuclear weapons would become increasingly stigmatised and delegitimised, their utility would diminish, and the focus would shift to non-nuclear alternatives.

Maintaining the Distinctness of Nuclear Weapons

During the US war in Korea, the US Administration had considered the use of nuclear weapons. However, an officer in the State Department's Bureau of Far East Affairs had then warned in 1950 that even though "military results achieved by atomic bombardment may be identical to those attained by conventional weapons, the effect on world opinion will be vastly different."[22] This statement embodies two very important points. First, that it is possible to wreak damage of comparable proportions by conventional weaponry as by nuclear warheads. In fact, this capability of conventional arms has only improved over the years and it handsomely matches the levels of lethality and destruction that small nuclear weapons are capable of. Fuel-air explosives, for example, are a new generation of highly lethal conventional weapons that are capable of delivering a blast similar to a small nuclear explosion over an area several miles wide.[23] This threatens to blur the distinction between the two categories of weapons. And this is more likely to happen if the US were to follow the suggestions put forth in the NPR to develop mini-nukes or nuclear warheads with a small yield.

The second and more important point, however, that can be discerned from the statement is that the strategic effect of the use of nuclear weapons is certain to be far more dramatic than that of conventional armoury, notwithstanding the largely similar damage to life and property caused in both cases. It is for this reason that the maintenance of the distinction between the nuclear and conventional weapon is extremely important to retain the nuclear taboo. Once a nuclear weapon has been used again, and the norm has been broken, then the risk of not only nuclear proliferation, but also nuclear use would rise manifold. Hence, the need to place nuclear weapons in a separate category as absolute weapons, restricting their role to purely nuclear deterrence and raising the argument of their use being in violation of moral and ethical human values. Nuclear weapons have acquired a special status that makes them the weapons of last resort and this should be encouraged. A nuclear weapon is not just another weapon of war, but comes higher up in the hierarchy as an extreme and unacceptable form of weaponry. An easy or first use of nuclear weapons must not be allowed to become a possibility by fostering a tendency to club them in the same category as other conventional weapons.

Accepting No-First Use (NFU)

The adoption of NFU would be a crucial step towards the eventual delegitimisation of nuclear weapons since it would involve an assurance from every NWS that it would not be the first to introduce nuclear weapons into a conflict. Since there will be no first use, it would effectively mean no use of the nuclear weapon. If NFU could be implemented through a legally binding, universally accepted agreement, then any first use would constitute a violation of international law, inviting crippling universal sanctions, worldwide opprobrium and any other penalties that could be consensually agreed to.

Of course, there are critics of the no-first use who dismiss it as nothing more than a declaratory policy that would mean little once hostilities actually break out between two nuclear nations. Such criticism, however, tends to overlook two facts. First, that the adoption of NFU automatically translates into a different nuclear force posture and deployment pattern that ensure that the promise of no-first use is kept. Therefore, a meaningful no-first use would have to be accompanied by the withdrawal or non-deployment of nuclear weapons that have war-fighting rather than deterrent functions. In fact, a pledge to negotiate the elimination of tactical nuclear weapons would give additional credibility to NFU.

The second salience of NFU arises from the accepted relevance of declarations on the conduct of war. These principles set certain limits to the levels of permissible violence and establish useful norms. For instance, the Geneva Protocol of 1925 outlawed the use of chemical and biological weapons and except for rare instance, the weapons have remained out of the realm of war-fighting. In fact, eventually, though several decades down the line, it did become possible to capitalise upon the norm to formalise conventional outlawing of these WMD.

Following this example, an NFU commitment from all the NWS could translate into a nuclear weapons convention some time in the future. At present, only two countries—India and China—accept NFU. The latter, however, does not offer an unconditional NFU,[24] in contrast to the one spelt out by India that is clear and unambiguous. Enshrined in the draft Indian nuclear doctrine that was put forth in August 1999, it categorically commits India to never be the first to use nuclear weapons. In doing so, it has opened up an alternative to the prevailing conventional wisdom of first use as espoused by the Western nuclear powers, and now, even Russia. It offers a counter-view to the traditional aggressive and arms race generating doctrines and could engender a new belief system. If all the NWS were to accept this principle, then none would ever initiate a nuclear strike. Naturally, therefore, over the years, the utility of nuclear weapons would decline.

Banning the Use or Threat of Use of Nuclear Weapons

A step even beyond NFU, an agreement prohibiting the very use or threat of use of nuclear weapons would be a meaningful step to delegitimise the weapon. In case all the NWS were to make a solemn declaration that nuclear weapons shall not be used and that any country using them or threatening to use them shall face commensurate retribution and a total boycott by all the countries of the world, it would make these WMD significantly impotent and useless. The value of nuclear weapons would fall instantly and further proliferation would voluntarily stop. None would want to acquire weapons that could not be used, not in war, and hence not as a deterrent either. Consequently, the unique status that nuclear weapons are deemed to provide, would no longer seem worth aspiring for. Meanwhile, even rogue states would no longer have any use for these weapons for fear of serious reprisals. Therefore, a total ban on the use of nuclear weapons would directly strike at the very roots of their utility.

Interestingly, the UN General Assembly has periodically considered resolutions to this effect. As far back as in 1961, it had adopted a declaration by a vote of 55 to 20 with 26 abstentions stating that the use of nuclear weapons was contrary to the "spirit, letter and aims of the UN". The US and NATO had then opposed it contending that in the event of aggression, the attacked nation should be free to take whatever action with whatever weapons not specifically banned by international law. India, for several years now, has proposed a "Convention on the Prohibition of Use of Nuclear Weapons" (UNGA 55/34G [L.30]).[25] It seeks a multilateral, universal and binding agreement prohibiting the use or threat of use of nuclear weapons through an international convention. In the year 2000, however, the move was able to muster only 109 votes in favour with 43 against and 16 abstentions. Predictably, the P-5 opposed the resolution. The US did so since it did not consider it a practical approach to the ultimate goal of nuclear disarmament. It hailed instead a step-by-step process that embraced unilateral, bilateral and multilateral measures.[26] Ironically, Japan, for all its abhorrence of nuclear weapons, also found it prudent to abstain for reasons similar to those voiced by the US. Such country positions, nevertheless, underestimate the significance of actually arriving at a convention banning nuclear weapons use to devalue them substantially.

Nuclear Reductions

Common sense dictates that reductions in nuclear arsenals should be a significant move towards promoting confidence building, averting the risk of nuclear war and, eventually, reducing the need for nuclear weapons. Given that in the context of the US and Russia, a deliberate nuclear attack is considered implausible, there can be no reason for retaining the kind of numbers that they continue to keep. Rather, the presence of such large numbers increases risks by raising the possibility of "accidental" or unintended nuclear war, especially since Russia's nuclear command and control systems or early warning mechanisms are not believed to be in the best of conditions. Smaller arsenals, then, would impose less strain on these mechanisms, besides being easier to safeguard and protect from accident, theft or unauthorised use. Moreover, arms reduction by the NWS would help to shore up global support for the non-proliferation regime and thereby contribute to greater international confidence building.

This should logically be the outcome of nuclear reductions. Unfortunately, though, not all nuclear reductions may lead to such positive developments as has become clear after the recent US pronouncements. Early in 2002, the US magnanimously announced its resolve to cut back its nuclear arsenal in the coming decade to a number as low as 1,700. But the decision is accompanied by several such steps that take away from its significance as a move towards devaluing nuclear weapons. In fact, as has already been discussed in the chapter on nuclear reductions, the latest US initiative ends up adding more value to nuclear weapons than otherwise. For one, it raises the levels of secrecy in the actual number of nuclear warheads that shall remain in the nuclear arsenal. The US has decided that the weapons removed from the active stockpile shall not be destroyed but shifted into a reserve force, the exact numbers of which are not known. This effectively means removing weapons from a transparent, deployed force to an opaque, non-deployed state from where they can be recalled if the president so desires. Therefore, the Bush Administration is cleverly adding to the "hedge," a reserve of thousands of nuclear warheads permitted by arms control treaties since they have mandated only the destruction of launchers and not warheads, while reducing "operationally deployed strategic nuclear warheads."

In this light, it becomes important to add a supplementary to reducing nuclear weapons. Mere reductions cannot be good enough. These must be done in good faith and to genuinely cut down numbers without raising the value of those that do exist. This calls for greater transparency in reductions. All nuclear warheads, regardless of their type, function, stage of assembly, associated delivery vehicles, etc. should be counted. Limits on the total inventory of nuclear warheads would minimise the reversibility of reductions and diminish the possibility of rapid breakout. Such limits would force the eventual dismantling of thousands of additional warheads, and improve the stability of nuclear balance. Suitable arrangements for exact, verified accounting and assured physical security of all warheads and fissionable materials and a mechanism for exchanging and verifying information about the location and status of warheads would also add to confidence levels in the reductions and shore up the case for non-proliferation.

De-alerting

In assessing the risks associated with nuclear arsenals, the operational and technical readiness of nuclear weapons for use is as important as the number of delivery vehicles or warheads. Elimination of continuous-alert status should, therefore, be pursued as a meaningful objective. This would reduce the perceived danger of short warning-time for attacks, provide greater chances of detecting preparations to use nuclear weapons and thereby increase the time available for crafting political solutions, and reduce pressures on command-and-control systems to stand ready to respond quickly, thereby decreasing the chances of erroneous or unauthorised launch of nuclear weapons. Above all, de-alerting would enhance the political relationships by eliminating the assumption that the other side might launch a surprise attack.

In fact, reductions in arsenals should be accompanied by a conscious effort at reducing alert levels. This is important because once the arsenals are smaller, there is a greater likelihood of countries with first use doctrines keeping them on hair-trigger alert so as to defy a "use them or lose them" syndrome. This could raise the risk of mistaken or unauthorised launch. The way out then lies in de-alerting in order to extend the time needed to prepare the weapons for a launch. De-alerting, or as some have called it, the process of "strategic escrow"[27] could be done through the physical separation of warheads from delivery vehicles or through the deployment of delivery systems out of the range of targets. If undertaken on a reciprocal basis, it would reinforce stability and reduce risks of an inadvertent launch.

Meanwhile, lengthening the fuse on their possible use would be a step towards downgrading the importance of nuclear weapons since military planners would have to discount their role in contingency planning. Over a period of time then, with a minimum number of nuclear weapons and no nuclear weapons ready to launch, the weapons could fall into a pattern of disuse and lose their utility.

Halting Further R&D for Nuclear Modernisation

Besides reducing the existing nuclear arsenals, it is also imperative that a complete halt be put to further modernisation of nuclear weapons. This is particularly important given the emergence of a new trend towards improving the weapons in such a way as to make them "more user-friendly". In the US, certain laboratories are reportedly pushing for the development of mini-nukes of smaller yields. In fact, in a little-publicised move, the US Congress lifted the ban on designing of new low-yield weapons in the year 2000.[28] The US justifies this by pointing towards threats arising from terrorism, and proliferation of smaller WMD capabilities to new states. It is contended that smaller, low-yield nuclear weapons could be effectively used to destroy even a deeply buried or hardened underground facility without causing too much collateral damage.[29]

Whatever the rationale, a decision validating the importance of smaller, low-yield and more usable nuclear weapons could set off a dangerous chain reaction with global ramifications. Russia is already believed to be thinking along similar lines. Victor Mikhailov, the former head of Minatom, has written in an article examining the role of nuclear weapons in the 21st century, "Of great importance [to Russia] will be [the future] development of a new generation of super-precise nuclear weapons of super-small yield and with a small impact on the environment."[30] The development of such usable weapons would signify a major reversal in devaluing nuclear weapons and, hence, should be opposed at every level.

Restricting Delivery Systems

Long range delivery systems like ballistic missiles and strategic bombers expand the role and scope of nuclear weapons. It, therefore, would make eminent sense to put in place some kind of universal restriction on delivery systems in a bid to devalue nuclear weapons. If the weapon is not effectively deliverable, then it would be of little use and, hence, not worth aspiring for or stockpiling.

In this context, it is suggested that the INF Treaty, presently in force only for the US and Russia, should be expanded to prohibit deployment anywhere of land-based missiles between the ranges of 500 and 5,500 km. In fact, to do so would be even more pertinent in the present context of the US NMD. The defence shield has raised the hackles of various countries, including US allies, and has further raised the danger of proliferation. A multilateralisation of the INF Treaty, meanwhile, offers another way of addressing the threat of errant missiles from small states, since most of them would fall within the range that shall be prohibited by a universalised INF Treaty.

Investing Confidence in Arms Control and Non-Proliferation Treaties

Nuclear arms control and non-proliferation initiatives have periodically been pursued to reduce the armaments of the NWS and to check further proliferation. In order to ensure that these measures are able to fulfill their objectives, it is necessary that countries exhibit commitment to, and faith in, them. Actions demonstrating a trend to the contrary, naturally, take away from the effectiveness of such treaties or agreements and could shake the faith of others as well. Several such instances from contemporary times could be mentioned in this context to illustrate the impact they could have from the point of view of proliferation and stability. The lack of progress in negotiating new formal arms control agreements, and in bringing already signed agreements into force, has contributed to a lack of confidence in using formal treaties to achieve key policy goals relating to non-proliferation and strategic arms reductions. In particular, the difficulties, protracted delays, and the current stalemate encountered in trying to bring the START II Treaty into force, the Senate's rejection of the CTBT, the resistance to passing enabling legislation for the CWC, the US withdrawal from the ABM Treaty, its reservations on the Kyoto Protocol, as also on the Protocol of the BWC, and the expressed desire for greater flexibility in adapting US strategic forces to the changing international environment have all contributed to a negation of arms control. It reveals an underlying lack of faith on the part of the US in the ability of these measures to guarantee its national security. Expectedly, other countries too would feel the need to reconsider their postures.

The treatment meted out to the multilateral treaties sends a signal to others—whether these should be seen as sacrosanct measures or as pieces of paper that could be amended or abrogated as and when interests deem it necessary, without paying adequate attention to the global ramifications of the same? Moves such as those being made by the US generate international instability, lower confidence levels and increase the value of nuclear weapons.

Adequate Verification and Enforcement

One of the most important arguments raised against nuclear abolition revolves around the twin problems of agreeing upon adequate verification procedures, and the infeasibility of ensuring the absence of cheating by a country. Lack of adequate mechanisms, institutional as well as technological, to oversee the process of disarmament has made the countries cling to their arsenals.

Therefore, in order to devalue nuclear weapons, it is necessary to put in place effective verification and enforcement strategies that clearly lay down penalties for cheating or breakout. The guarantee of a collected and concerted response could act as an effective deterrent to the clandestine development of nuclear weapons once all countries have committed to their non-use or abolition. The importance of broad international support in this regard is best demonstrated through the example of Iraq. In the early 1990s, when the clandestine nuclear weapons programme in Iraq was first discovered, it led to an international outcry that resulted in the consensual establishment of the UNSCOM (United Nations Special Commission). However, over the years, as aspersions came to be cast upon the preponderant US role and objectives in the UNSCOM, the consensus has broken down, and in 2002, no nation other than the UK has come out in support of US strategy to militarily seek the removal of Saddam Hussein.

The case of North Korea provides another example of how a country is deemed to have used its pursuit of a nuclear weapons programme to extract concessions from countries that felt threatened. In its negotiations with the US in 1994 that led to the Agreed Framework, Pyongyang agreed to halt its nuclear weapons programme in exchange for certain rather lucrative concessions from the US and others. This generates a sense that the country has not only got off rather lightly for its misdemeanour but has rather been rewarded. Such feelings promote the temptation to go down the nuclear path in order to use it as a lucrative bargaining chip. In order to stem this trend, it becomes necessary that sure penalties be guaranteed for any breach of trust. Only a collective endeavour to respond surely and in a united manner could effectively serve to devalue nuclear weapons. Of course, a determination of the exact nature of penalties that may be acceptable to all requires further investigation. But it is not a challenge that cannot be surmounted if the requisite political will exists.

Conclusion

The chapter has identified several steps that could help to devalue the importance of nuclear weapons. These are largely based on common sense and logic. However, country positions are not always dictated by these two parameters. Ample evidence of this is provided by national positions as revealed in the voting patterns on the different resolutions that continue to be placed before the UNGA on the issue of disarmament. In the year 2000 (as also every year on an average) nearly four dozen such resolutions appeared before the UN. However, an examination of the voting patterns revealed the rather predictable position of each country. The NWS and their allies tend to support resolutions listing the progress already made in the direction in terms of the increasing membership of the NPT, the conclusion of the CTBT, supposedly ongoing negotiations on the FMCT, reductions negotiated under the START process, etc. However, they oppose all resolutions seeking an actual delineation of steps to devalue and eliminate nuclear weapons in a time-bound or phased manner.

Not surprisingly, the two most relevant resolutions introduced by India,[31] year after year, and containing most of the steps outlined above have been dubbed as "unrealistic" and unacceptable for failing to take note of the progress made unilaterally and bilaterally in the direction of arms reductions and other non-proliferation measures. However, India has consistently maintained that while non-proliferation is a positive step in the direction of containment of nuclear weapons, the basic argument of the NPT and the discriminatory treatment meted out to nations has an endemic syllogistic fallacy. The pursuit of non-proliferation without an adequate movement towards discarding nuclear weapons from the arsenals of all is not a sustainable objective.

In conclusion, it may be reiterated that devaluing nuclear weapons could be an effective way to ultimately achieve nuclear abolition. This is relevant because complete elimination has often been dismissed as unattainable unless war as a means of settling disputes between states is first abolished and a state of general and complete disarmament is enforced through a system of world governance. Since the attainment of these prerequisites appears Utopian, nuclear disarmament too tends to be dismissed as unattainable. But simple and easily implementable steps to devalue nuclear weapons prove that there are worthwhile ways of arriving at nuclear abolition. In this context, it is heartening to draw upon the experiences of outlawing of other classes of weapons and activities without waiting for a comprehensive solution. In fact, it must be realised that nuclear weapons are in a class by themselves, considering the indiscriminate, comprehensive, and long-lasting destruction they can cause. For the sake of the future of mankind, then, it is imperative that these steps be taken.

While it is the nation-state, through appropriate legislation, that can be the only effective implementing authority of these steps, the ambience for doing so can be created through a build-up of public awareness and opinion. During the 1960s and 1970s, civil society had manifest itself as a powerful tool in anti-nuclear peace movements. Unfortunately, though, such activism faded out without achieving the desired results. While it its beyond the scope of this chapter to examine the reasons for their failure,[32] in the present context, it shall be important for the rational elements in every society to impress upon their governments the need to implement the steps to devalue nuclear weapons so that all may be freed of the fear of a nuclear war.

Notes

1. Much of this optimism was emanating from a reassessment of threats on the part of the US and Russia after the collapse of the Soviet Union. The predominant view among American security analysts of the time was that Russia lacked the capacity or even the intention to threaten or attack Western territories. CIA Director Robert Gates even told the US Congress in 1992 that "the threat to the US of deliberate attack from the [former Soviet Union] has all but disappeared for the foreseeable future." The then Russian foreign minister echoed the very same sentiments when he stated in the same year that this nation "no longer views the US as a foe." Tom A. Zamora, "Put A Safety Cap on Testing", *Bulletin of Atomic Scientists,* March 1992, p. 25.
2. The spirit of the changing times was aptly caught by an international analyst in these words, "These are the days of hope, not despair...It is time to strip away the complex and arcane strategic theory of the Cold War and start from scratch". See Lincoln Wolfenstein, "End Nuclear Addiction," *Bulletin of Atomic Scientists,* May 1991. In a similar mood, the Stimson Centre came out with a publication in the mid-1990s that argued that the character of international relations "is undergoing an irreversible transformation that would eventually invalidate rationales for weapons of mass destruction." This report, prepared by Barry Blechman and Cathleen Fisher, is mentioned in Robert A. Manning, "The Nuclear Age: The Next Chapter," *Foreign Policy,* Winter 1997-98, p. 76. Also see William Arkin, Damian Durrant and Hans

Kristensen, "Nuclear Weapons Headed for the Trash," *Bulletin of Atomic Scientists,* December 1991, p. 16.

3. "An American Legacy: Building a Nuclear-Weapon-Free-World", The Final Report of the Steering Committee Project on Eliminating Weapons of Mass Destruction (Washington D.C.: Henry L. Stimson Centre, March 1997).

4. Annual Report of the US Department of Defence as quoted in Office of the Secretary of Defence, *Nuclear Weapons Systems Sustainment Programme* (Washington DC: DoD, May 1997).

5. Neo-realists support the retention of nuclear weapons capabilities on the ground that they exercise a fundamental restraining influence on inter-state conflict. For more on this, see John J. Mearsheimer, "Back to the Future," *International Security,* vol. 15, no. 1, Summer 1990; Kenneth Waltz, "The Spread of Nuclear Weapons: More May be Better", *Adelphi Paper No. 171* (London: International Institute for Strategic Studies, 1981); Charles L. Glaser, "The Flawed Case for Nuclear Disarmament," *Survival,* vol. 40, no. 1, Spring 1998, pp. 112-128. Glaser argues that disarmament is likely to increase the probability of nuclear war between disarmed states since cheating by a nation to build these weapons would always be a possibility.

6. It can be argued that modern day conventional weapons can cause a similar extent of destruction. But the difference lies in the fact that to do so would require a more intensive and extensive use of air power. Nuclear weapons, meanwhile, offer the temptation of easy use and, therefore, also invoke the fear of irresponsible use.

7. For a detailed description of the kind and scale of destruction and damage a nuclear attack could cause, see Jonathan Schell, *The Fate of Earth* (London: Picador, 1982).

8. Jonathan Schell, "The Folly of Arms Control," *Foreign Affairs,* Autumn 2000.

9. The inadequacy of export controls has been proved very often. So much so that post-September 11 and despite official reports of the dangers of nuclear terrorism that should have tightened security measures at all levels of nuclear access, Pakistan has still been reported to have succeeded in secretly buying from Britain special high grade aluminium for its uranium enrichment plant in Kahuta. A 47 tonne consignment of the high strength aluminium is known to be used in making centrifuges for converting uranium ore to bomb-grade uranium 235. "Pak Shops for K-Kit in Britain," *The Times of India* August 26, 2002.

10. Glaser, n. 5, p. 115.

11. James Schlesinger, former director of the Central Intelligence Agency of the USA, as quoted in Ken Booth and Nicholas J. Wheeler, "Beyond Nuclearisation," in Regina Cowen Karp, ed., *Security Without Nuclear Weapons? Different Perspectives on Non-Nuclear Security* (Stockholm: SIPRI, 1986).

12. Preamble of the UNESCO Constitution.

13. Jerome B. Wiesner, "A Militarised Society," in Len Ackland and Steve McGuire, eds., *Assessing the Nuclear Age* (Chicago: Educational Foundation for Nuclear Science, 1986), p. 227.

14. Statement made by Under Secretary Walter Slocombe in his testimony made before the International Security, Proliferation and Federal Services Sub-Committee of the Senate Government Affairs Committee on February 12, 1997. He is quoted in the report produced by the Office of the Secretary of Defence entitled "Nuclear Weapons Systems Sustainment Programmes (Washington DC: DoD, 1997).

15. Statement made by Col. General Yakolev, commander-in-chief of Russia's Strategic Missile Forces.

16. "Russia's Controversial New National Security Concept," *Disarmament Diplomacy,* issue no. 43, February 2000, http://www.acronym.org

17. "The Alliance's Strategic Concept," NATO Press Release, NAC-S(99) 65, April 24, 1999. http://www.nats.int/docu/pf/1999.

18. "National Security Concept of the Russian Federation: January 2000," as reproduced in *Strategic Digest,* March 2000, p. 297.

19. Robert G. Joseph, "Nuclear Deterrence and Regional Proliferators," *Washington Quarterly,* vol. 20, no. 3, Summer 1997, p. 168.

20. Speech delivered by French Prime Minister Alain Juppe at the *Institut des Hautes Etudes de Defence Nationale,* Paris on September 7, 1995 and as reproduced in *Strategic Digest,* April 1998.

21. Wolfgang K.H. Panofsky and George Bunn, "The Doctrine of Nuclear Weapon States and the Future of Non-Proliferation," *Arms Control Today,* vol. 24, no. 6, July/August 1994.

22. Nina Tannenwald, "The Nuclear Taboo: The United States and the Normative Basis of Nuclear Non-Use," *International Organisation,* vol. 53, no. 3, Summer 1999, p. 444.

23. Ibid., p. 460.

24. China maintains that its NFU commitment does not apply to its own territory or territories that it claims as its own. Hence, the ambiguity regarding the possibility of use of Chinese nuclear weapons in a conflict over Taiwan, or certain portions of Arunachal Pradesh, an Indian state to which China lays claim.

25. For the complete text of the resolution see http://www.un.org/news

26. See "Appendix: Summary of Resolutions," *Disarmament Diplomacy* issue no. 53, December 2000, http://www.acronym.org

27. Stansfield Turner, "The Specter of Nuclear Proliferation," *Security Dialogue,* vol. 29, no. 3, 1998, pp. 293-301.

28. Robert Alvarez, "A Long Season of Discontent," *Bulletin of Atomic Scientists,* January/February 2001.

29. Theresa Hitchens, "The US Nuclear Debate: Issues of Concern," BASIC Paper no. 35, February 2001.

30. Ibid., p. 57.

31. "Reducing Nuclear Dangers" and "Convention on the Prohibition of Use of Nuclear Weapons." For text, see *UN Disarmament Yearbook,* vol. 24, 1999.

32. For more on this aspect, see Lawrence S. Wittner, *One World Or None: A History of the World Nuclear Disarmament Movement Through 1953,* vol. I and II (Stanford, Calif.: Standford University Press, 1993) and Stuart Croft, *Strategy of Arms Control: A History and Typology* (Manchester: Manchester University Press, 1996).

10

Future of Nuclear Deterrence: Credibility and Consequences

Jasjit Singh

Although it has existed throughout the history of human conflict, deterrence as a subject was extensively explored in a new context after nuclear weapons came into existence. But we often forget that deterrence is not an objective by itself—it is the means to achieve the objective. By its very definition, a country may adopt one of the many options in terms of the means and methods available as long as it can make the use of nuclear weapons credible. This would, to a significant extent, depend upon the objective sought to be achieved. Some states have sought to find in nuclear weapons the solution to deter, and if required, defeat, far superior conventional forces. Others believe that nuclear weapons could prevent war itself because of the risk of total annihilation. And yet others set the objective of deterrence to stop the threat and use of nuclear weapons by another country against themselves.

Deterrence itself has been based on two dimensions: through denial, and by punishment. The former holds out the prospect of denying the adversary success or victory at costs that a country believes to be less than the advantages of waging war. On the other hand, deterrence through punishment holds out the promise of grievous hurt and damage independent of the denial dimension. Historically, deterrence had essentially functioned in its denial dimension since the means to impose punishment independent of fighting a war were not available. Naval gunboat diplomacy started to bring that element into a deterrent equation although it remained applicable to countries with significant coast lines and/or navies. It was the advent of air power at the beginning of the 20th century that brought in a capability to target and destroy human-economic-military assets of the adversary directly without having to conduct a war on the ground. The ability of air power to transcend physical barriers and military forces on the ground provided a capability that could impose punishment through destruction on the adversary without having to engage in a war on the surface. Missiles and nuclear weapons raised the dimension level of deterrence to new heights.

It is not surprising, therefore, that the use, usability and utility of nuclear weapons have been constructed around the logic of deterrence, even where the use is perceived to go beyond what would logically constitute deterrence. But it is surprising that it continues in practice to lead to confusing conclusions. For example, very often, the use of threat of force to compel a certain shift in the target country's policy choices is also included in the concept of deterrence. On the other hand, deterrence as a concept and strategy has existed from times immemorial. It is not intended here to go into the theory and practice of deterrence, but rather, to examine the trends that would influence nuclear deterrence in the future except in so far as it requires some references to the conceptual and theoretical aspects of deterrence.

Trends Affecting Deterrence

There are a number of trends that are affecting the way nuclear deterrence would evolve in the coming years and these are briefly discussed below:

- While nuclear proliferation had been placed in a constrictive strait-jacket by 1995 when the NPT was extended indefinitely, and the regime itself expanded and deepened, proliferation risks and incentives have, if anything, increased due to the policies being followed by the US and other leading powers. There is no movement towards abolition of nuclear weapons and even notional attention to achieving such a goal seems to be disappearing. The war against Iraq to denuclearise it (when no evidence of the existence of weapons of mass destruction, nor any sign of nuclear weapons or missiles to hit Israel leave alone the US, was found or had really existed even before the war), and the hesitation in dealing even with kid-gloves with North Korea which has for all practical purposes claimed possession of nuclear weapons (also acknowledged by senior US leadership) clearly send a message that possession of even rudimentary nuclear weapons would provide sufficient deterrent capability to even a weak power against a superpower threatening to use force against it. This is likely to provide a strong incentive to many countries in the future to seek nuclear deterrent capabilities.

- The vulnerability of modern states and society to violence, especially to the threat and/or use of nuclear weapons, has been increasing markedly due to many factors, independently of the level and quality of military power available to strike them. For example, globalisation, industrialisation and urbanisation are progressively increasing the interdependence of different parts of society even in the developing world. A great deal of daily life in modern conglomerate centres of population depends heavily on the efficient functioning of food and water supplies, medical aid, communications, and a host of other utilities and services even from beyond the border of the country. The terrorist attacks on September 11, 2001, in New York and Washington, highlight the vulnerability of modern states to terrorism even through unarmed airliners used as a weapon of mass destruction. The increasing vulnerability of states and society also implies that stakes in deterrence and attack would be correspondingly higher. Greater destruction and disruption now can be achieved with lesser quantities of destructive power, especially since it can be delivered more precisely. Nuclear deterrence would, therefore, be achievable at progressively lower levels in the future as compared to those of earlier periods.
- Long-range strikes with an array of conventional weapons (like the thermobaric weapons) with high destructive power and platforms capable of hitting targets with a high degree of accuracy under adverse weather and visibility conditions are already a reality. At the same time, smaller/micro-yield nuclear weapons with highly accurate delivery at longer distances even in adverse weather, on one side, and expanded effects of conventional weapons, on the other, are reducing the distinction between nuclear weapons and conventional arms. This provides capabilities that were visualised in the formulation of the concept of "Discriminate Deterrence" over 15 years ago.[1] In substance, this process would make nuclear weapons more usable by countries that have such capabilities, profoundly altering the nuclear deterrence paradigms.

- The end of the Cold War has also put an end to the strategic framework for the stability and arms control paradigm in place for more than four decades. Strategic stability had been crafted in a bipolar context by the two superpowers with other allied countries being brought into the equations as part of that bipolarity. With the collapse of one superpower and its changed relationship with the other, the basic framework of the strategic environment has fundamentally altered. Strategic stability now has to be sought in an increasingly multilateral and multinodal framework. As regards arms control, the same phenomenon has affected its key parameters. Arms control for strategic stability now is required to be crafted in a far more difficult to achieve multilateral context where the needs of stability and the goals of arms control in different states would be different and difficult to reconcile in a multilateral context. It is not surprising, therefore, that many of the multilateral regimes proposed for implementation, like the CTBT, FMCT, etc. which had overflowed from the Cold War period, lie dormant in near-coma, or, like the bilateral ABM Treaty, have regressed. The trend, in fact, is in the reverse direction where unilateralism to pursue national interests on the strength of military power, even through preemption, is taking root in the national security strategy, not only of the US but also of lesser powers, though their ability to do so would remain constrained because of the limits of the capacity and ability of their power to do so.
- Introduction of missile defences promises to bring in fundamental changes in the nuclear deterrence paradigm from exploitation of the vulnerability of the target country with no defence, to exploitation of superiority for offensive use concurrently with a (hopefully superior) defensive capability to neutralise hostile nuclear threats delivered by ballistic missiles. While this may promise enhanced security for the country that is able to deploy a credible effective defensive shield, the overall impact would be to reduce stability, increase the propensity to preempt, and introduce an element of higher level of risk in case of failure of defence for whatever reason. Compellence strategies would tend to assume greater utility for political purposes than classical deterrence.

Deterrence or Compellence?

Belief systems constitute the "software" that drives practical policies as well as the acquisition and maintenance of the "hardware" of capabilities. On the other hand, all major changes in human history have come about due to a change in belief systems. Abolition of slavery, apartheid, colonialism, etc. are vivid examples. Nowhere do belief systems influence policy more than in the ideas for the use and usability of force; and this is taken to apocryphal levels by nuclear weapons. They also play a critical role on issues of security, stability and risks. These belief systems, often perpetuated due to reasons dependent on cultural correlates or elite interests, evolve into principles and precepts for the use and utility of nuclear weapons. These principles, in turn, manifest themselves in the doctrines guiding strategy and postures. As in the case of conventional military doctrine, these could be offence-oriented or defensive. In other words, nuclear doctrines are an important indicator for making an assessment of the scope for security/stability or otherwise and the dynamics of deterrence.

The fact that there is no credible defence against nuclear weapons in a way makes it easier to assess the impact of nuclear doctrine on stability. If we are to use nuclear doctrines of deterrence as the core of assessment for stability and/or defence, then the following issues/questions amongst many others must be addressed:

(a) What is the central rationale for nuclear weapons articulated by the country as seen in its explicit or implicit doctrine for the employment of nuclear weapons? Does the doctrine define the purpose and aim of nuclear weapons precisely or is substantive ambiguity built into the doctrine? The more narrowly defined the goals and objectives of keeping nuclear weapons, the easier it would be to assess their impact on stability. Conversely, if the doctrine is kept ambiguous and/or seeks to employ nuclear weapons across a broad spectrum of scenarios and contingencies, the greater are the prospects of strategic instability.

(b) Is the nuclear doctrine supportive of offensive-aggressive goals and intentions or does it direct defensive strategies? Doctrines of "first use," leave alone the concepts of preemption and launch on warning, have been at the root of tremendous insecurity. Over 50 incidents have been recorded where nuclear weapons were threatened to be used in one form or the other. It is notable that except for one incident (where China is reported to have conveyed a threat to the US in 1996) all the other threats were held out by nuclear weapon states pursuing nuclear doctrines which relied on first use and essentially pursued an offensive strategy.

(c) What are the conditions under which nuclear weapons would be employed? Would they be tightly controlled by a civil political authority or by dictatorial military regimes lacking broad-based decision-making legitimacy, and intrinsic checks and balances? Would hair-trigger alert status strengthen security and stability? What are the risks of unauthorised (including accidental) use and miscalculations?

(d) What is the relationship between the nuclear doctrine of a country and its philosophy for the use of force? The history of warfare indicates that many countries have relied on ever increasing levels of targetting and destruction of civil society wherein nuclear weapons had become an almost inevitable progression by the middle of the 20th century. MAD was justified during the Cold War. But the Cold War ended more than a decade ago; and the erstwhile protagonists are now either members of the same military alliance or "partners for peace." But have the doctrines altered in consonance with these changes?

(e) What is the relationship or balance between offensive-defensive doctrines of likely adversaries? Any asymmetry would tend to weaken strategic stability. But so would symmetrical offensive-oriented doctrines since they are likely to reduce the margin of error. The prospects of strategic stability will be higher in case both sides pursue defensive doctrines and strategies. But in reality this does not happen and, hence, the risks and dangers associated with nuclear weapons increase, especially in terms of crisis instability.

(f) Above all, we need to keep in mind the reality that nuclear weapons seek to provide security by generating insecurity and exploiting vulnerability at their extreme level. This at best is a fragile basis even if necessitated by nuclearisation of the security environment. Therefore, the question: does the doctrine encourage perpetuation of nuclear weapons or support their abolition?

Nuclear strategy bordering on theology not only came to project a military role for nuclear weapons, but also supported the doctrines and strategy of nuclear war-fighting. It took four decades of nuclear weapon deployment before the heads of the two superpowers came to what should have been the obvious conclusion right from the beginning: that nuclear war cannot be won, and, therefore, must never be fought. But if we look beyond the superficial, it has been clear that nuclear weapons were more an instrument of politics rather than a military instrument of war-fighting. The end of the Cold War, by ending the militarised confrontation, has, if anything, made this aspect even more clear.

Mahatma Gandhi, with his deeper understanding of human nature and the affairs of men and states, had recognised from the very beginning that the atom bomb played a political role. Contrary to what might have been expected of the apostle of peace, he did not react for months, at least publicly, to the use of the atom bomb in Japan in 1945. However, his concern about its nature, especially as a horrendous weapon of mass destruction, was intense. He was to later describe it as a diabolical instrument of violence. But for months after the bombs were dropped on Hiroshima and Nagasaki, he kept silent because his concern was about the possible "use" of the atomic bomb to deny India its independence. It was only after the commitment to India's independence became unambiguous that he came out against the bomb publicly.

Subsequent history only helped to confirm the nature of the role that nuclear weapons actually play. Compared to one occasion of actual use (at Hiroshima and Nagasaki) in 1945, there have been at least 51 identifiable incidents where nuclear weapons and forces were alerted for use or were threatened to be used. (See Table 10.1). Compared to the threat implicitly emanating from the very existence and possession of nuclear weapons, these incidents represented threats that were explicit, although at varying levels.

There are a number of characteristics common to these incidents. The key characteristic has been that of the factor of the existence of asymmetry of nuclear weapons capability. The threatened country either had no nuclear weapons or a significantly lower capability to retaliate against the state that chose to convey the threat. In many cases, the parties were not necessarily at war. In one case, it can be argued that the "use" was against a friendly country to compel it to cooperate. From a closer study of the actual incidents, it is possible to conclude that favourable asymmetry was a factor encouraging the holding out of the nuclear threat.

For example, there were a number of threats held out by the US against China during the 1950s. Some of them are believed to have been conveyed through India since there was no direct channel of communication between the US and China. But there is no evidence of threats being held out after 1964 when China acquired nuclear weapons. On the other hand, as the range and size of the Chinese nuclear arsenal increased, the US recognised the government in Beijing, something that it had refused to do more than two decades before that, and actually sought a close relationship in establishing a new strategic triangle. The centrality of the ideological goal of fighting Communism was set aside and nuclear weapons came to play a different political-diplomatic role than the one they had till the mid-1960s.

Table 10.1: **Incidents Involving Threat of Use of Nuclear Weapons**

Incident		By Whom	Date
1.	Iran	USA	1946
2.	US aircraft shot down by Yugoslavia	USA	November 1946
3.	Inauguration of president in Uruguay	USA	February 1947
4.	Security of Berlin	USA	January 1948
5.	Security of Berlin	USA	April 1948
6.	Security of Berlin	USA	June 1948
7.	Korean War: Security of Europe	USA	July 1950
8.	Korean War: Entry of Chinese troops	USA	November 1950
9.	Korean War: To compel Chinese acceptance of a ceasefire	USA	April/May 1953
10.	Security of Japan/South Korea	USA	August 1953
11.	Gautemala accepts Soviet Bloc support	USA	May 1954
12.	Vietnam: Siege of Dien Bien Phu	USA	July 1954
13.	China-Taiwan Conflict: Tachen Islands	USA	August 1954
14.	Quemoy	USA	1955
15.	Suez Crisis	USA	October 1956
16.	Suez Crisis: To compel withdrawal of British and French troops from Suez	USSR	October 1956

17.	In connection with the Berlin crisis	USSR	1957-58
18.	Political crisis in Lebanon	USA	July 1958
19.	Political crisis in Jordan	USA	July 1958
20.	China-Taiwan conflict: Quemoy and Matsu	USA	July 1958
21.	Security of Berlin	USA	May 1959
22.	Laos	USA	1961
23.	Security of Berlin	USA	June 1961
24.	Berlin crisis	USSR	1961
25.	Emplacement of missiles in Cuba	USSR	September 1962
26.	Soviet emplacement of missiles in Cuba	USA	October 1962
27.	Withdrawal of US missiles from Turkey	USA	April 1963
28.	Confrontation between Indonesia and Malaysia	UK	April 1963
29.	In connection with China's nuclear weapon programme	USSR	1964
30.	To compel the termination of Israel's offensive on the Golan Heights	USSR	June 1967
31.	Pueblo seized by North Korea	USA	January 1968
32.	Vietnam: Siege of Khe Sanh	USA	1968
33.	To compel termination of Chinese initiated incident on the Sino-Soviet border	USSR	1969
34.	Vietnam War	USA	1969-72
35.	Indo-Pak War 1971[1]	USA	December 1971
36.	Bangladesh War[2]	USA	December 1971
37.	Vietnam Negotiations	USA	1972
38.	Arab-Israel War	Israel	October 1973
39.	Arab-Israel War	USA	October 1973
40.	Arab-Israel War	USSR	October 1973
41.	Security of Iran	USA	1980
42.	South Atlantic War	UK	May 1982
43.	Military Exercise (Brasstacks)	Pakistan	January 1987

44.	Kashmir Crisis	Pakistan	April-May 1990
45.	Gulf War	USA	January 1991
46.	India-Pakistan	Pakistan	1994
47.	In connection with Taiwan (elections)	China	March 1996
48.	Kargil War	Pakistan	May 1999
49.	India-Pakistan tensions	Pakistan	May 2002
50.	Korean Crisis	North Korea	2002-03

Notes: 1. Against Soviet Union; 2. Sailing of *USS Enterprise.*

Sources: Data based on Barry Blechman (*Force Without War)* and other documents; Daniel Elsberg, interview with *New York Conservation Press,* 1980; Richard Nixon, *Time* Magazine, July 29, 1985; Kargil Review Committee (Sage, 1999), author's data base.

In each of these incidents, nuclear weapons played an important political role rather than a military one although the military factor was visible in the demonstration. The threatened party could ignore the threat only at its peril. The adverse asymmetry, therefore, became an important factor in shaping the responses of the country which was put under pressure.

Deterrence Strategies

The subject of nuclear deterrence has been examined at length during the past six decades and it appears almost superfluous to cover the beaten track. But, as in many other things, a brief re-look at the well-known is also helpful to focus our minds on the issues at hand. To arrive at a clearer understanding of nuclear deterrence, it may be useful to dig into history and briefly examine the strategies adopted by different nuclear weapon powers over the decades.

For more than a decade after the US used nuclear weapons in August 1945, its military establishment tended to view the weapons as an extension of conventional weapons for strategic bombing, except that these would be less demanding on resources of aircraft and air crew, and more devastating in effect. There was no strategy in the immediate post-War period for the specific use of nuclear weapons. It was the civilian strategic-security community that started to define the basic elements of strategy for the new weapons of mass destruction. Some people may argue that the original concept of the weapons as a war-fighting instrument was never really abandoned over the following decades. How much of the initial sentiment was guided by the horrors and costs of World War II may remain uncertain. But the elements of nuclear deterrence were clear. Herman Kahn, Bernard Brodie, and others started to announce the advent of deterrence and concretise the concept. For example, Bernard Brodie wrote in 1946, "Thus far, the chief purpose of our military establishment has been to win wars. From now on, its chief purpose must be to avert them. It can have almost no other purpose."

For nearly a decade and a half, US nuclear strategy was conditioned by near total preeminence of the US nuclear arsenal in spite of the USSR having tested its first nuclear weapon in 1949. This was largely due to the fact that this did not automatically translate into an effective nuclear strike capability against the US. But by 1954, when the USSR tested its hydrogen bomb, the evolving threat was clear and John Foster Dulles announced the doctrine of "massive retaliation" to meet the perceived Soviet challenge. US strategy, especially as seen in its targetting policies of the time, seems to have relied heavily on counter-force strategy as a continuation of its belief that nuclear weapons were a usable military instrument with high military utility rather than the broader idea of deterrence.

By the time John F. Kennedy came to power in 1961, the idea of credible deterrence had taken sufficient root to ensure that nuclear strategy would be guided by such thinking. The USSR had started to acquire the capability to target the US homeland with nuclear weapons. Robert McNamara, as the defence secretary, began to emphasise sufficiency, credibility and, above all, survivability to enable a "second strike" capability as intrinsic to any successful nuclear strategy for the United States. With growing Soviet nuclear and conventional military capabilities and the need to ensure Alliance credibility, extended deterrence assumed great importance. By June 1962, McNamara had refined his strategy where "the principal military objectives, in the event of a nuclear war stemming from a major attack on the (NATO) Alliance, should be the destruction of the enemy's military forces, not of his civilian population." But the Cuban missile crisis had an undoubted impact on US strategy.

The US came to the conclusion that the level of punishment that a nation is willing to bear rather than abstract numbers of nuclear weapons or the size, throw-weight ratios, etc. of the nuclear arsenal would be critical to the success of deterrence. But in addition to what it believed necessary for counter-force employment of nuclear weapons, the US continued to build its nuclear arsenal for counter-value strategy, going as far as to arrive at force level needs based on empirical calculations of what would be needed to destroy Soviet society and industry to make it a non-viable nation-state. The world came closest to a nuclear exchange during the Cuban missle crisis in late 1962. In a curious divergence from the lessons of the Cuban crisis, by 1967, McNamara was to declare, "It is our ability to destroy an attacker as a viable 20th century nation that provides the deterrent, not our ability to partially limit damage to ourselves." How much of this was due to Alliance strategies or internal turf goals (where the Energy Department built warheads and the Pentagon promoted delivery systems) remains uncertain. But the fact was that the US had been sufficiently deterred in 1962 from launching even a conventional strike against Soviet missiles and nuclear weapons in Cuba in spite of its overwhelming capacity to wipe out the Soviet Union with nuclear weapons, and opt for a negotiated settlement essentially because its military could not ensure that no hostile nuclear weapon would get through. The complete destruction of the Soviet Union was simply not worth the loss of even a single American city.

The strategy of MAD as it evolved, thus, was conditioned by different sets of realities and paradoxes. By the end of the 1970s, the United States acknowledged that it could no longer credibly deter most hostile actions by the threat of nuclear retaliation, and that nuclear weapons could usefully serve the primary purpose of deterrence against either nuclear weapons or overwhelming non-nuclear attacks. Essentially, deterrence strategy relied on exploiting the vulnerability of the adversary to very high levels of destruction against which it would have no credible defence. The search for reliable defence finally led to the development of missile defences to back up the offensive strategy on which the nuclear weapon doctrine was based. In many ways, the US has been shifting back to the original thinking (which it never really gave up) of nuclear weapons as another form of war-fighting system rather than one of deterrence in specific circumstances. The perceived superiority in technological capabilities no doubt provides a powerful argument to seek unilateral solutions and reduce the complications of extended deterrence strategies. In substance, the US is moving toward a strategy that relies not only on punishment, but also on the denial dimension of deterrence.

The Soviet Union, on the other hand, had remained focussed on deterrence through punishment, and that too on a massive scale, recognising that even a conventional war would rapidly escalate to the nuclear level; and a tactical use of nuclear weapons would quickly lead to a strategic exchange. To that extent, Soviet nuclear strategy, as it evolved over time, completely discounted the NATO concept of an escalating limited nuclear war. So the Soviets concentrated on building heavy strategic forces and expanded into tactical nuclear weapons to back up their large conventional forces when NATO deployed similar capabilities. But they continued to believe in nuclear weapons as an essential part of the preparations for war. Deterrence from that perspective rested on the denial strategy, based on the demonstrated capability and willingness to fight a nuclear war, and, hence, deny a victory to the enemy, acting as the deterrent to aggression. There was a view that this was inevitable in an overall strategically defensive posture, but which the West still interpreted as an aggressive offensive strategy. The Cuban missile crisis showed, however, that Moscow was as concerned with exercising restraint beyond a point as Washington was.

While the Soviets believed in nuclear deterrence through denial, they were deeply disturbed by the concept of the SDI, or Star Wars, which promised to transform nuclear deterrence strategies if technology could be harnessed successfully. Since the Soviets continued to develop SDI itself and sought to curb deployment rather than development, it is not clear whether their concern was more related to the gap between the ability of the US to deploy SDI well before the USSR could, or because of a concern that the introduction of defensive capabilities complicated nuclear strategies. Considering that Russia accepted the abandonment of the ABM Treaty with significant equanimity, it would appear that the problem had more to do with the capabilities gap rather than a doctrinal disconnect.

The British approach to nuclear weapons initially was no different from that of the US. Its military saw the nuclear weapon as simply one more type of weapon for strategic bombing. At the end of World War II, the Chiefs of Staff were pressing for rapid build-up of nuclear weapons on the grounds that they would need large numbers, as indeed they did of other weapons. The Chief of Air Staff even put in a demand through normal procurement channels for an atomic bomb in August 1946! By 1952, Britain decided to place greater reliance on nuclear weapons for its defence in relation to perceived threats from the Soviet Union. But it was the 1957 Defence White Paper that laid down the policy that large conventional forces would no longer be necessary and the nuclear deterrent would be the mainstay of British defence. This was in line with the NATO strategy at that time which stipulated that a full-scale Soviet attack could not be repelled without resort to a massive nuclear bombing of the Soviet sources of power. From then on the British nuclear deterrent has been a part of the NATO collective deterrence strategy and force structure. After the end of the Cold War, Britain has retained only the nuclear-powered submarine-based nuclear deterrent.

French nuclear strategy reflected both the Gaullist world view and the position of France in it and a practical nuclear doctrine that sought to create a third decision centre. The independent deterrence was sought to provide "proportional deterrence" within the context of the Soviet threat. In essence, it recognised that the proportionate comparison of nuclear forces was not helpful in judging the deterrence needs once a minimum capability had been reached. Proportional deterrence, therefore, would be established once a country had the means to mortally wound the aggressor, if determined to do so, and when the aggressor is convinced of it, regardless of the proportion of weapons held by either side. The crux of deterrence was not parity or equivalence in numbers and throw-weight ratios that seemed to mark the US-USSR deterrence equation, but a sufficient number of nuclear weapons being adequate, regardless of the adversary's arsenal, as long as this was able to achieve that "mortal" quality. French nuclear policy clearly contains the element of using nuclear weapons capability as a political and foreign policy tool. It sought a greater role for France by proposing a tripartite control of NATO strategy on the grounds of its independent deterrent. Failing to achieve its goal, it continued to maintain an independent deterrent while cooperating with NATO as and when the security situation demanded it.

In spite of key differences in national approaches, there remained a substantive commonality of nuclear strategy among the industrialised developed powers. It was China that headed a new approach to nuclear strategy as soon as it acquired nuclear weapons. On the day following the nuclear test in October 1964, Chinese Premier Chou Enlai wrote to heads of governments stating: "The Chinese Government consistently stands for the complete prohibition and thorough destruction of nuclear weapons. China has been compelled to conduct nuclear testing and develop nuclear weapons. China's mastering of nuclear weapons is entirely for defence and for protecting the Chinese people from the US threat. The Chinese Government solemnly declares that at no time and in no circumstances will China be the first to use nuclear weapons."[2] The "no-first-use" (NFU) strategy sustained since then is in sharp contrast to those of the US, USSR/Russia, France and the UK, which have not given up the first-use doctrine even after the Cold War ended and the new partnership for peace programmes have been in place for years.

China's nuclear arsenal and deployment continue to be consistent with the declared doctrine of NFU. But it has now defined its other commitment not to use nuclear weapons against a non-nuclear state by the NPT rule for non-nuclear weapon status. In recent years, two other changes have been noticeable in its nuclear strategy. One is the substantive military modernisation programme of the past two decades that includes significant upgradation of the nuclear arsenal also. China is shifting to mobile land-based solid-fuel missiles for nuclear delivery, bringing greater substance to its strategy since it would enhance the survivability of the nuclear arsenal. Expansion of the space programme also strengthens the overall nuclear capabilities, especially for early warning and retaliatory strikes. There has been some speculation about the implication for the doctrine and strategy of the terminology of "limited deterrence" being increasingly used by the Chinese in recent years.

More important, China is believed to have conveyed a nuclear threat to senior US officials in 1996 on the issue of Taiwan and its possible move toward independence. It also deployed its forces and fired a number of ballistic missiles abeam Taiwan to back up its threat. Since then, there have been reports of Chinese concentration of ballistic missiles in hundreds being deployed abeam Taiwan. These developments are hardly consistent with the doctrine and strategy of NFU of nuclear weapons held out by China since 1964.

Israel clearly has acquired nuclear weapons for deterrence of the larger Arab conventional military forces. This naturally implies a first use strategy even if weapons are not deployed on immediate alert. Pakistan has also justified its acquisition of nuclear weapons on the grounds of India's conventional superiority that, it believes, hangs like the sword of Damocles over it.[3] Senior generals dealing with nuclear policy have spelt out four scenarios for the use of nuclear weapons: a force threshold, space threshold, economic threshold and the survival of Pakistan. The essence of its nuclear strategy is similar to that of Israel and NATO: to deter a superior conventional military force from over-running the country. The difference is that Pakistan had started to use sub-conventional war since 1983 under the nuclear umbrella on the assumption that the low-cost covert war by itself would not constitute sufficient provocation for escalation to a full-scale conventional war. And if it does, the risk of further escalation to nuclear levels would be sufficient to provoke the international community to intervene, thus, placing Kashmir on the international agenda. In fact, all the instances of nuclear threat by Pakistan appear to have been based on such assumptions. Thus, the strategy is a mix of political use of nuclear weapons with sub-conventional war to achieve political ends.

India, on the other hand, adopted a nuclear strategy quite different from that of the earlier nuclear weapon states who planned for the use of such weapons merely as more destructive bombs for strategic bombing which would only provide greater capability for war-fighting rather than change anything basic. It would take a long time for them to understand and accept the central role of nuclear weapons which makes war with them un-winnable because both contestants would be destroyed under the MAD situation. Nuclear weapons do not possess any military utility since they cannot serve a rational purpose in fighting a war that destroys both the victor and the vanquished. But because of their immense destructive power and non-availability to all but a few, their value for political compellence and a coercive role is almost unlimited. India had the benefit of learning from the experience of others and avoiding their mistakes. It adopted a nuclear strategy not different from that of China: of deterring nuclear weapons threat and use. It adopted a strategy of NFU and a posture which can be best described as one of "recessed deterrence." But more of it in Chapter 11.

Table 10.2. **Spectrum of Nuclear Strategy Options**

Deterrence	Capabilities and Consequences
Maximal	— Doctrine of first-use. — Aggressive, offensive strategy, exploiting the vulnerability of the target country to the fullest and promising an end to the state itself.

	— Deterrence against conventional forces as well as nuclear weapons. — Second strike critical to credibility. — Full-range of nuclear capabilities. — Hair-trigger alerted arsenal (even on warning). — Very high risk of accidental launch, miscalculation and uncontrolled escalation. — War-fighting doctrine, with no direct political or military utility.
Minimum	— Sufficient deterrence at the minimal level of arsenal to impose unacceptable levels of destruction. — Strategically defensive while relying on offensive action to cope with deterrence failure. — Second strike key to credibility. — Deterrence against nuclear weapons as well as conventional weapons, separately and collectively depending upon national strategies. — Less complex command and control. — Deployed arsenal, fully alerted and ready to launch at very short notice, or even in a first strike preemptive strategy if deterrence is believed to be failing.
Recessed	— No-first-use doctrine and strategy. — Deterrence against nuclear weapons only. — Strategically and tactically defensive. — De-alerted, non-deployed, de-mated weapon systems which can be brought together for use in a short time. — Little or no risk of accidental launch or through miscalculation. — Robust command and control but with far less complexity and size than other strategies. — Escalation controllable, more manageable. — Survivability for punitive retaliatory strike critical to credibility (like second strike).
Extended	— Military security alliance-based quasi-nuclear weapon status, without legal ownership. — First use, even if in the "last resort" by the

	guarantor nuclear weapon state would be critical to credibility of extended deterrence. — Adoption of a no-first-use doctrine by the primary nuclear guarantor would nullify credibility of extended deterrence. — Alliance dependency narrows options.
Existential	— Possession of nuclear weapons itself provides a level of deterrence capabilities independent of doctrine, strategy or posture. — Factor in foreign and security policy of states dealing with country possessing nuclear weapons. — Constraints, even if limited, on security and military policies of other countries.
No deterrence	— Non-nuclear status. — Autonomy and security curtailed and circumscribed by the interests and policies of nuclear weapon states.

There is no standard template for nuclear deterrence and it has taken different shapes with different countries and on different occasions. The US and the USSR adopted a maximalist approach to nuclear deterrence, relying on massive retaliation with tens of thousands of nuclear weapons deployed across the world on hair-trigger alert. The UK and France adopted a more modest approach, but still adopted the doctrine of massive retaliation along with NATO commitments. China adopted a fundamentally different approach with the minimum deterrence strategy, relying in addition on an NFU doctrine. As a consequence, its arsenal did not rise above a few hundred even when it was fully developed. Cost was obviously not the reason for this policy since continuation of modernisation of the arsenal that it possessed was more expensive than building more warheads and delivery systems of existing types since the costs of research and development would already have been absorbed.

The history of nuclear deterrence is based on the assumption of certainty that deterrence would work any time and every time. But unlike conventional deterrence, the risks of any infirmity or failure of nuclear deterrence would lead to apocalyptical consequences. And yet, whichever way it is examined, nuclear deterrence intrinsically contains many elements of uncertainty. And this uncertainty increases markedly during periods of crisis when the risk of irrationality also increases significantly as the time available is very short. It is vital, therefore, that any deterrence strategy adopted must maximise the effect of deterrence while minimising the uncertainties and ambiguities implicit in any nuclear strategy, especially under circumstances where deterrence has failed and conflict emerges.

The deterrence theory relies on three criteria for its success: capability, communication and credibility. The threat underpinning deterrence is that aggression would almost inevitably result in retaliation with nuclear weapons. If that threat lacks credibility, then deterrence would be uncertain at the minimum, and meaningless at the maximum. While some would argue that ambiguity would produce a more effective posture, the credibility of deterrence would depend heavily on the certainty of response and the rationality of the decision-makers. The main problem, however, is that the adversary may not act in a wise and rational manner as expected. We must, therefore, be ready to accept that nuclear deterrence can fail to work in a situation of crisis and consequent uncertainty and misjudgement of the adversary's intentions. It needs to be noted that where critical decisions about nuclear weapons use may be taken by any one person in isolation to others or within a very small coterie (like in Pakistan in 1970-71) within the power structure or by ignoring the likely actions of the adversary, the risks of failure or irrational response could be significantly higher.

Corporate decisions that evolve through some sort of a process of corporate thinking and assessment, on the other hand, would inevitably produce a more balanced and rational approach to nuclear deterrence issues provided such decisions rely on accurate, reliable and timely intelligence. The Iraq War has highlighted the risks and weaknesses of intelligence assessments that go into critical military decisions. The dangers of such errors of intelligence and assessment are obvious in scenarios where decisions about the likely use of nuclear weapons are concerned. Such decisions have to be made under situations of extreme stress and in an emotionally charged climate. Nuclear weapons launch, especially against another nuclear weapon state, requires the most reliable and unambiguous intelligence about an adversary's capabilities, intentions and actions if unintended catastrophe is to be avoided. The highest level of certainty, of course, flows from the actual use of nuclear weapons by an adversary since this is not liable to any questionable interpretation. The credibility of one's deterrent strategy under these circumstances would rest on sufficient nuclear arsenal (including redundancy in command and control) surviving for an effective punitive retaliation after being hit with nuclear weapons. Such a punitive strike would have to ensure that it results in what any rational person would consider as an "unacceptable" level of punishment. This itself points to the need for a counter-value strategy.

The Asian Scene

It is against this background that we need to assess the increasing nuclearisation in the Asian security environment. Today, eight (Russia, China, India, Pakistan, Israel, France which has territories in the Indian Ocean, and the US which is fully deployed in Asia and the contiguous oceans with its nuclear-armed forces, and North Korea) out of the nine nuclear weapon states are in Asia. The ninth nuclear weapon state, the UK, has long-range nuclear-powered nuclear-armed submarines that can be easily deployed in the Asia hemisphere. Nuclear weapons have been used so far only once (on two cities) but on an Asian country. Of the 50 incidents of where threat of use of nuclear weapons can be identified (beyond the existential threat), 38 involved Asian countries or a crisis related to countries in Asia. More important, the US and Soviet weapons have been a reality for more than half a century. There has been no movement toward abolition of nuclear weapons even more than a decade after the Cold War ended, in spite of treaty obligations. The expansion of Western military alliances under the SEATO and CENTO in the late 1950s brought the nuclearised Cold War to Asia. Nuclearisation of the Asian security environment intensified in the early 1960s when China acquired nuclear weapons, and nuclear-armed naval forces of the two superpowers started being deployed in the Indian Ocean. China and the Soviet Union fought a bitter border war in 1969, in what would go down in history as the first direct war between two nuclear neighbours during which nuclear threats were held out.

The fact that a major territorial dispute existed between China and India where the former had illegally occupied 48,000 sq. km. of Indian territory and claimed another 94,000 sq. km. while absorbing 10,000 sq. km. illegally transferred by Pakistan, ensured that India could not ignore the reality of Chinese nuclear weapons. During the past two decades, China has been expanding the capabilities of its nuclear arsenal and has been modernising it across the board. China also changed its earlier posture of expressing a willingness to enter into nuclear arms reduction agreements once the superpower arsenals reduced to half their peak levels (of the 1980s). But as the Cold War started to wind down and agreements between the two superpowers to reduce their arsenals by 50 per cent were being concluded, China regressed from its earlier position and now insisted that it would join any arms reduction process only after the two superpowers came down to levels equal to those of China. Since the US and Russia are firm that they will retain their nuclear weapons in sizeable numbers for an indefinite period, it is clear that, notwithstanding its demands for total elimination of nuclear weapons, China will also retain its nuclear weapons for an indefinite period. The existential threat from these nuclear weapons would, therefore, continue for an indefinite period into the future.

The expanding presence of the superpower naval forces in the Indian Ocean had progressively increased the quality and quantity of nuclear weapons deployed to the south of India. The implications of this development were harshly demonstrated when the US despatched a powerful naval task force led by the mega-carrier USS *Enterprise* to the Bay of Bengal during December 1971 with the specific purpose of coercing New Delhi and making it respond in accordance with US policy. The task force was armed with nuclear weapons. That the Soviet Navy that stalked this force was also armed with nuclear weapons only emphasised the role of nuclear weapons.

India had been worried about a possible threat or demonstration of force by China, especially if the deteriorating situation in East Pakistan/Bangladesh erupted into uncontrollable mayhem. The Soviet Union had been seeking a treaty of peace and friendship for some time before 1971. But it was only in August 1971 that New Delhi agreed to such a treaty and went further by incorporating a clause on security. It is generally believed that this clause was meant to be a political deterrent to China so that it did not intervene. But strangely the intervention finally came not from China, but from the US and that too with nuclear and military coercion. It is also relevant to note that President Richard Nixon had stated much later in an interview in 1985 that the US would have used nuclear weapons, ostensibly to deter China from threatening India.[4] But the lesson that nuclear weapon states could attempt threats and coercion was clear.

India did not have to invoke the security clause in the Indo-Soviet

Treaty. But the lessons were clear. There is also no doubt that the 1974 PNE was what it was claimed to be. But it also sent a clear signal to Moscow that India would not need to rely on the security clause of the bilateral treaty. This clause, in fact, was dropped when the treaty was renewed in 1992.

But it was under the pressure of this increasing nuclearisation that India's nuclear policy underwent a fundamental change. The very nature of the nuclear weapons and India's policy of non-alignment implied that India would need to rely on its own nuclear deterrent if it was to successfully safeguard its own interests. But the decision was to exercise the maximum restraint, and defer the acquisition of the weapons to a later date. Meanwhile, all elements required for weaponisation at a future date were to be built up. Thus, was born the policy and strategy of keeping the option open. The issue of India's nuclear doctrine is covered in some detail in Chapter 11.

There is also a concern that a doctrine of no-first-use would expose India (and its population) to a first strike by an adversary. *Prima facie* this is true to a large extent. But what needs to be seen clearly is that this danger has existed all along, and certainly ever since the Indian security environment was nuclearised. Theoretically, the choice of the use of nuclear weapons against India by China has existed since 1964 and by Pakistan since 1987 although the political-strategic rationale for embarking on such a course did not exist. But Pakistan has held out nuclear threats a number of times since then. Any objective analysis of the US despatch of the nuclear-armed Task Force 74 headed by the USS *Enterprise* would indicate that the action carried an implicit nuclear weapons threat.[5] If no such actual use took place, it was entirely due to their own choice rather than any deterrent factor. The issue, therefore, is that a nuclear deterrent would reduce the incentive for possible threat of use or use of nuclear weapons against India by any country. But the doctrine of NFU does place India at risk to a first strike by another country.

The crux of the issue here, therefore, is not that the risk of a first strike against India would exist but what would be India's capability to reduce if not eliminate that risk. Without possession of nuclear weapons, there would be no capability at all to deter a nuclear weapons threat. With nuclear weapons, there would be two broad options: that of a first preemptive strike against an adversary, or an NFU strategy. India has opted for the latter. In the former case, the risks and dangers are much higher in spite of the fact that a much larger arsenal on hair-trigger alert would be needed. The potential for miscalculation and accidents increases exponentially under these circumstances. In fact, this was the condition under which the world existed during the East-West confrontation which also led to the build-up of such huge arsenals. This was also the major reason for the extensive deployments of a variety of weapons and delivery systems, extended deterrence, and extensive elaborate global command and control systems. This strategy had also led to concepts and capabilities of launch-on-warning, etc. Obviously, this approach would demand tremendous resources without any certainty that it would protect India from a nuclear attack.

It is in this context that the concept of NFU must be seen. The adversary would be able to undertake a first strike; and this is more likely to seek decapitation of the arsenal rather than attempt a counter-value strike. But this is exactly why survivability of the nuclear arsenal is so critical. Sufficient force must survive to undertake punitive retaliation which, in the words of the doctrine, would inflict "unacceptable damage" on the aggressor. This would aim to provide the necessary deterrent against a first use by the adversary. The aggressor will have to take into account that sufficient arsenal would survive to inflict a punitive retaliation which would be at a level it would find unacceptable. Here it must also be noted that increasing vulnerabilities of modern states and greater interdependence in a globalised world are already reducing the level of force required to deter another country, especially with nuclear weapons.

The risks of nuclear threat and possible use increase with the prospects of an inter-state war where one or both sides possess nuclear weapons. A conflict situation, therefore, lowers the nuclear threshold by increasing the risk of possible use. India's interests are better served if there is no nuclear threat. This is why abolition of nuclear weapons is perceived as a national security objective. In the absence of nuclear disarmament, there is a need to keep the nuclear threshold as high as possible so that the nuclear threat factor does not enter into India's operational security situation. India's interests would be prevention, and avoidance of war would naturally raise the nuclear threshold. It would also be necessary to keep conventional military capability sufficiently high to provide a credible effective deterrent against conventional war. This would help to raise the nuclear threshold. It can, of course, be argued that if India maintains a higher level of conventional military capability than before, this would increase the threat perceptions in Pakistan which may then tend to lean even more toward early reliance on nuclear weapons. This only enhances the importance of the credibility of Indian nuclear deterrence.

On the other hand, there is no objective reason for threat perceptions in Pakistan to increase since its need to reach out for the nuclear button could rationally arise only in the case of the success of an Indian invasion which it can no longer hold with its conventional forces. But Pakistani leaders will have to face up to the truth that all wars so far since 1947, including the latest one in the Ladakh sector of J&K state, were initiated by Pakistan; and all that Pakistan has to do, therefore, is to not start another war. But by the same logic, the reverse would be true in the case of China which in any case is modernising and upgrading its military power in a significant way.

Some Concluding Thoughts

We can draw some broad conclusions from the discussion so far. First, every state evolves its own paradigm and doctrine of deterrence depending upon a number of factors. For example, the capabilities of the state would have an over-riding influence on the deterrence strategy that is possible and might be adopted. Similarly, the nature of the threat that is required to be deterred adds to that parameter. At the same time, the state concerned has to come to a judgement about the aim of its deterrence capability. In other words, does it seek to deter war itself and, hence, the use of conventional forces, or deter the threat and use of nuclear weapons and nothing else? Many other imponderables like strategic culture and perceptions of the ruling elite also start to impinge on the type of deterrence capability and the way it might be used by a state.

Second, the history of nuclear deterrence over the decades clearly establishes that in the face of the adversary also possessing nuclear weapons, even if it is at a significantly smaller scale, the credibility of deterrence would rest on a second strike capability. In other words, the ability to retaliate with nuclear weapons after absorbing a nuclear attack actually defines the ability to deter successfully. If second strike is the touchstone of credible deterrence, then logic would dictate that it should be possible to plan for a retaliatory strike after the enemy has struck (which he can always do as a surprise attack) without investing in a first strike capability. This, in essence, is the doctrine adopted by China and India which intrinsically lowers the risks and dangers of miscalculations, accidental launch and unauthorised launch, besides requiring a robust, but far less complex, command and control system. However, this is feasible essentially where the country pursues a strategically defensive policy.

Third, Napoleon's dictum that God is on the side of the bigger battalions has been rendered completely obsolete by nuclear weapons. Unlike conventional forces, numbers are relevant only to the extent of their utility in causing what the adversary may believe is an "unacceptable" level of destruction. This level may vary from country to country. Massive destruction would be possible by even one nuclear bomb. There would remain a degree of ambiguity about the level that the adversary may judge as unacceptable to it. But, in principle, the numbers do not have to match those of the adversary. China is a clear example which created, what from all accounts was a credible deterrent against the USSR and the USA, combined with a very small arsenal.

This leads us to the fourth conclusion, that the key to credible nuclear deterrence lies in its punishment dimension. Fighting a war with nuclear weapons (like with other weapons) would not provide any scope for winning against another nuclear-armed state since the war would lose all rational purpose in the massive indiscriminate annihilation of both sides. Any assumption of being able to control the process of use of nuclear weapons in some sort of graduated employment is flawed at the very least. In fact, any exchange of nuclear weapons would inevitably function in the punishment domain. This is why mutual vulnerability was perceived as a critical basis of successful deterrence. The current trend toward bringing in the denial dimension to the nuclear doctrine through ballistic missile defences promises to complicate the overall situation with negative effects.

Fifth, any use of nuclear weapons, on or beyond the battlefield or theatre of war, would inevitably have a strategic impact. The terminology of "tactical" nuclear weapons may have served some political-diplomatic utility, especially in the context of the extended deterrence of the Cold War, but as a strategy it did not in any way provide a different level of effect in political-strategic terms. It is interesting that the bombs dropped on Hiroshima and Nagasaki would get classified in today's world as "tactical," in terms of the delivery platform as well as the warhead. But no one would obviously argue that they were not strategic in their impact.

Last, but not the least, nuclear weapon capabilities have made wars between nuclear weapon states far more dangerous than among those who do not possess them. Nuclear weapon states, therefore, have tended to adopt strategies of indirect approach to apply force and violence for political and ideological ends. During the Cold War, the two heavily nuclear-armed adversaries fought hundreds of wars through proxy, harnessing ideology wherever possible, including that of *jihad* adopted by the most powerful liberal secular democracy, the US, to fight the "non-believer" Soviet Union in Afghanistan. The sheer existence of nuclear weapons with both adversaries imposes major limitations on the way force and violence can be used against each other without risking a nuclear exchange. This alters the very nature of war.

Notes

1. Fred C. Ikle and Albert Wohlstetter, *et. al., Discriminate Deterrence,* Report of the Commission on Integrated Long-Term Strategy, January 1988, p. 2.
2. Timothy Garden, *Can Deterrence Last? Peace Through a Nuclear Strategy,* (London: Buchan & Enright, 1984).
3. Former Foreign Minister of Pakistan Agha Shahi, address as chief guest, in Fasahat H. Syed, ed., *Nuclear Disarmament and Conventional Arms Control Including Light Weapons* (Rawalpindi: FRIENDS, 1997), p. 421.
4. "A Nation Coming Into Its Own," *Time* Magazine, July 29, 1985, p. 27.
5. In fact, then US President Richard Nixon later stated that he had considered using nuclear weapons in 1971. See *Time* Magazine, July 29, 1985.

11

Nuclear Doctrine for Defence

Jasjit Singh

China and India decided to adopt a nuclear doctrine different from that of any of the existing nuclear weapon states. When China tested its first nuclear weapon on October 16, 1964, it also announced that it would not be the first to use nuclear weapons at any time and in any circumstances. It has continued this position as a formal declaration since then.[1] However, this was not given any serious consideration during the following decades. In January 1994, China formally presented a draft for the Treaty on the No-First-Use of Nuclear Weapons to the USA, Russia, the UK, France and other countries, proposing that the five acknowledged nuclear weapon states hold first-round discussions on the proposed treaty in Beijing at the earliest. But this did not find favour with the other acknowledged weapon states.

India had supported the concept for a no-first-use (NFU) of nuclear weapons even when it had not acquired nuclear weapons. After it formally declared its nuclear weapon status on May 11, 1998, it also committed itself, like China, not to be the first to use nuclear weapons. This also implies non-use against non-nuclear weapon states, however that is defined. India, over the decades, in fact, had sought global elimination of nuclear weapons in order to denuclearise its own security environment substantively because such an environment would intrinsically provide it with a competitive advantage in conventional military capabilities although it was never officially argued from that perspective. The draft nuclear doctrine was prepared by the National Security Advisory Board and released by the government as a public document on August 17, 1999. Subsequently, the government formally announced the broad details of the doctrine which unambiguously specifies a no-first-use nuclear strategy for India. In recent years, the German Government had asked NATO to adopt a no-first-use doctrine without success. Thus, China and India are the only two countries in the world today to follow a no-first-use doctrine.

Even at the risk of stating the obvious and the risk of repetition, one must start any discussion of nuclear doctrine and strategy with some of the basic issues that impinge on deterrence and nuclear weapons strategy. This is all the more necessary since the concept of no-first-use has never been examined in any meaningful way although the two major nuclear weapon states of Asia have opted to follow this approach. To begin with, deterrence has been part of military strategy and political relations of states from the very beginning of human history. But it is important to look at the methods by which deterrence works. In substance, the final arbiter of deterrence is not just the rationality or the posture of deterrence, but the question of how it would function in circumstances when deterrence fails. In other words, the capabilities that come into play in case of deterrence failure are the ones that finally decide the credibility of deterrence in the first place rather than the mere existence of such capabilities.

Some Deterrence Issues

At the very beginning, we must note that the term "deterrence" is used very often rather loosely to cover a number of aims and objectives. For example, compellence and coercion are treated as deterrence. Compellence or coercion is obviously not compatible with an NFU posture since they are based on the use and threat of use of force first. To that extent, this aspect is not included in the present discussion. But a clear definition of deterrence issues is critical to any examination of strategy since it defines the goals and objectives of that strategy. Deterrence may operate through one of the following methods:

Denial,

Punishment, or

A combination of denial and punishment.

In theory, of course, a combination of the denial and punishment dimensions of deterrence can operate. But our concern here is in relation to nuclear weapons. The nuclear revolution has ensured that deterrence through denial by the use of nuclear weapons would not lead to raising the costs only on the adversary, but also on one's own self as long as the adversary also has nuclear weapons and is willing to use them. The impact of a nuclear revolution, therefore, becomes an important aspect that deserves attention in examining the strategy for credible deterrence. The main characteristics of the nuclear revolution may be summed up as follows:

- Enormously high destructive capabilities of nuclear weapons, especially against counter-value targets.
- Extremely high vulnerability of the state and society to the possible use of nuclear weapons. This vulnerability is increasing with the passage of time as the world moves deeper into globalisation and greater interdependence in every dimension of the modern industrialised life.
- No credible defence against the use of nuclear weapons, especially when delivered by ballistic missiles. While work is going on for the creation of missile defences, it is not certain how effective these would be and what consequences would flow from their deployment in various circumstances. Even if successful, they would provide defence against ballistic missiles only, and not necessarily against all other forms of delivery of nuclear weapons.
- Rational behaviour of states and their leaders in pursuance of their own vital interests forms the very basis of deterrence theory and practice.
- Unlike conventional military capabilities, mere numbers and balance of forces are not critical because of the scale of destruction that could be caused by even a single nuclear weapon is enormous. Napoleon's saying that God was on the side of the bigger battalions is no longer valid in the context of the nuclear revolution. This was also the central lesson of the Cuban Missile Crisis forty years ago.
- Nuclear deterrence credibility over the decades has come to be associated with the survivability of a minimum force for "second strike" retaliation. In other words, it is a second strike capability that provides the foundation of credibility to the nuclear deterrence strategy rather than a first strike capability.
- Risk of decapitation due to surprise attack is an integral component of the nuclear strategy.

- Risks and consequences of accidents and potential use of nuclear weapons through miscalculation and false triggers remain the biggest challenges which could result in catastrophic consequences.

NFU as a Political Statement or Strategy

Declarations of no-first-use are, of course, a political statement. But they can also represent the basis of nuclear strategy. There has always been some doubt in many quarters about whether China's NFU is a political statement or forms the basis of its nuclear strategy. Similarly, there are many people, including in India, who doubt the viability of a no-first-use nuclear strategy as propounded by the National Security Advisory Board and adopted by the government. This view tends to see the commitment that India would not use nuclear weapons first as simply a political statement.

NFU as a political statement is useful as a diplomatic-political tool to create an environment of confidence and lower the sense of threat. But it is claimed that under the pressure of a serious crisis situation where the vital interests of the country are at stake, the NFU commitment would not hold back the leadership from using nuclear weapons first if in their judgement this was necessary, or it provided a critical advantage. Others, like Pakistani elites, have dismissed the concept as posturing and merely an attempt to gain a moral high ground without substance. Realists also claim that an NFU pledge is neither reliable nor verifiable. Hence, its doctrinal utility is highly suspect. On the other hand, if NFU is pursued as a strategy, then a number of consequences becomes discernible even in an opaque arena.

Compared to a mere political statement, an NFU strategy would require an appropriate force structure and definitive posture to support it, and in spite of the secrecy that countries are bound to maintain on their nuclear capabilities, sufficient indications would be available for making a distinction between a political posture and a nuclear strategy. For example, there have been nearly 49 incidents since 1946 when nuclear weapons were threatened to be used in varying degree. Most of these can only fall under the strategy of compellence and coercion rather than deterrence. The reported nuclear threat by China to the US in 1996 on the issue of Taiwan, thus, would be incompatible with China's NFU pledge. One must also consider whether deployed nuclear missiles (like ICBMs) in silos) are compatible with an NFU strategy?

The spectrum of nuclear strategy options available to states for deterrence has been examined in Chapter 10. It may be noticed that "recessed deterrence" indicates another level of deterrence available to policy-makers. In a way, this is the strategy that India seems to be following although this is not articulated in these terms anywhere. In fact, the term had been rapidly confused with "non-weaponised deterrence" in the early 1990s in spite of critical differences between the two concepts.[2] Even a cursory study would indicate that compellence or coercive use of nuclear weapons would not be compatible with a recessed deterrence strategy and this naturally narrows the role of nuclear weapons.

Strategy Options

There is also a concern that a doctrine of no-first-use would expose the defender (and its population) to a first strike by an adversary. *Prima facie* this is true. But what needs to be seen clearly is that this danger has existed all along after the security environment was nuclearised. The issue is not that such exposure to nuclear attack exists, but that what defensive strategy would deter such an attack.

The crux of the issue here, therefore, is not that risk of a first strike against India would exist but what would be our capability to reduce, if not eliminate, that risk. Without possession of nuclear weapons, there would be no capability at all to deter a nuclear weapons threat. With nuclear weapons, there would be two broad options:

- A first, even preemptive, strike against an adversary, or
- A no-first-use strategy.

India has opted for the latter. The risks and dangers in the former case are much higher in spite of the fact that a much larger arsenal on hair-trigger alert would be needed. The potential for miscalculation and accidents increases exponentially under these circumstances. In fact, this was the condition under which the world existed during the East-West confrontation which also led to the build-up of such huge arsenals. This was also the major reason for the extensive deployments of a variety of weapons and delivery systems, extended deterrence, and extensive elaborate global command and control system. This strategy had also led to concepts and capabilities of launch-on-warning, launch-under-attack, etc. Obviously, this approach would demand tremendous resources without any certainty that it would protect a country from a nuclear attack.

It is in this context that the credibility of no-first-use must be seen. The attacker would be able to undertake a first strike; and this is more likely to seek decapitation of the arsenal rather than attempt a counter-value strike. But this is exactly why survivability of the nuclear arsenal is so critical. Sufficient force must survive the attacker's nuclear strike to undertake punitive retaliation which, in terms of strategy, would inflict "unacceptable damage" on the aggressor. This would aim to provide the necessary deterrence against a first use by the adversary. The aggressor will have to take into account that sufficient arsenal would survive to inflict a punitive retaliation which would be at a level it would find unacceptable. Here it also must be noted that increasing vulnerabilities of modern states and greater interdependence in a globalised world is already reducing the level of force required to deter another country, especially with nuclear weapons.

On the other hand, if the attacker has a huge arsenal, then it is possible to argue that it could completely overwhelm the defender country through a single massive strike. For example, the US or Russian capability, in theory, could produce such a result against India, Pakistan and possibly even China. But this flies in the face of any rational purpose of such strike. In such cases, it is also debatable whether a first strike strategy would produce the political results even in a preemptive mode.

Indian Doctrine

India became the sixth state with nuclear weapons in May 1998. The broad parameters of the nuclear policy were spelt out in the following months. On August 17, 1999, the government released the draft nuclear doctrine for India prepared by the National Security Advisory Board (NSAB) to the National Security Council (NSC) to encourage public debate. The document was considered by the government and the broad contours of the official doctrine announced in January 2003. For the purpose of the current study, it would be useful to focus on the proposal of the NSAB since it provides more details in elaboration of the core aspects of the doctrine announced in January which is presumably based on the 1999 proposal of the NSAB and was believed to have been triggered by many demands for switching to a first-use doctrine.

The proposed doctrine naturally attracted a great deal of comment which was the stated purpose of releasing it to the public. Unfortunately, many comments in India and abroad have confused the very concept of a "doctrine", ignoring the accepted meaning of the term and in spite of a clear expression in the Preamble of the paper where, in para 1.6, it is stated that "this document *outlines the broad principles* for the development, deployment and employment of India's nuclear forces" (emphasis added). A doctrine is expected to define the principles and concepts for the guidance of policy-makers after it has been approved by duly constituted authorities. It must not be confused with policy, strategy or even posture which no doubt may be expected to flow from it. This elementary fact seems to have been missed by most critics. For example, many people have complained that the doctrine does not indicate the "cost" of India's proposed doctrine. The issue of costs is very important. But should a doctrine paper go into costs? If it does, would it remain a document that outlines "broad principles" or become a policy planning paper? On the other hand, the document consistently states that India would pursue a doctrine of credible "minimum" deterrence. The concept of minimum is not merely related to a single aspect of, say, numbers, but also to costs and posture.

Similarly, there has been a view that the doctrine does not spell out the size of the arsenal, or for that matter, the likely threat. This was to be expected since the doctrine would hardly be expected to define these issues. The strategic and security environment is not static and keeps changing consequent to a range of factors, depending upon the actions and attitudes of other states. Advances in technology impact on strategic capabilities and military power in various ways. While national security interests would normally remain constant, the framework under which they have to be pursued are subject to change. Thus, the minimum level of capability required cannot remain static. For example, the needs of deterrence are likely to decay with the passage of time. But it would be difficult, and even imprudent to follow a policy based on making a firm judgement. The doctrine, therefore, provides a clear enunciation of the principle that the concept of credible minimum deterrence is a dynamic one where the actual size, components, deployment and employment of nuclear forces would be related to the changing realities of the strategic environment, technological imperatives and the needs of national security.

Much of the critique of the document has risen from two factors. The first is the rather unique approach of making the proposal public for discussion and debate. Unlike conventional weapons whose development, deployment and employment normally remain confined to military and defence establishments, nuclear weapons concern all citizens. In an established democracy like ours, people have a right to know which principles will govern nuclear policy. The second is even more important. The doctrine does not adopt the conventional wisdom of other nuclear weapon states. To that extent this is not only in contrast to the acknowledged wisdom of the main nuclear powers but seeks to chart a new path.

The central logic of the doctrine is built on five essential principles which are discussed in the following paragraphs.

First, India's nuclear weapons are meant *only to deter nuclear weapons threat/use.* It is obvious that nuclear weapons cannot possibly provide deterrence in all circumstances. For example, they are unlikely to deter sub-conventional wars (including terrorism) or even conventional wars. In fact, limited conventional wars remain a distinct possibility even where nuclear weapons are present. At best, nuclear weapons may deter full-scale conventional war that leads to a "decisive" military result or capture of large/valuable territory when the risk of escalation to nuclear weapons would be high. On the other hand, the experience of the nuclear age bears witness to the conclusion that limited war is still possible under the nuclear overhang. The Kargil War was not the first such war between two nuclear neighbours. What is often forgotten is that thirty years earlier China and the Soviet Union fought an intense local, limited war across the border on the Ussuri River. This is why the doctrine does not expect that nuclear weapons would deter all forms of war and armed conflict. But unlike most other nuclear weapon states, India's nuclear weapons are not meant to deter the use and threat of use of conventional weapons, or a generalised formulation of protecting national interests any time, anywhere. The doctrine requires that the nuclear policy should seek to deter rather than fight a war with nuclear weapons. The orientation of the doctrine is, thus, quite narrowly focussed on the need to defend India through deterrence against nuclear weapons threat.

Second, the doctrine rests on the principle of "no-first-use" even against nuclear threat or use. India's use of its nuclear weapons would be predicated on failure of deterrence, that is, if and when an adversary uses nuclear weapons against India. This also defines the parameters of potential development, deployment and employment of the nuclear arsenal. The doctrine commits Indian nuclear policy to *retaliation only.* As may be seen, para 2.4 of the NSAB doctrine paper expresses it in the following terms:

> The fundamental purpose of Indian nuclear weapons is to deter the use and threat of use of nuclear weapons by any State or entity against India and its forces. India will not be the first to initiate a nuclear strike, but will respond with punitive retaliation should deterrence fail.

The only country to adopt a similar commitment is China although uncertainties have always existed about whether China's commitment constitutes a political statement or is a doctrinal precept. It is also not clear whether China's commitment applies to the territories claimed by China as part of its country. A political statement would be reassuring but cannot be relied upon to be sustained in times of crisis. But a doctrine of no-first-use would also require a strategy and posture consistent with the doctrine. While the doctrine itself cannot be verified (as much as a political statement cannot be accepted as binding), the posture can certainly be judged as to its consistency with the doctrine. And any change in the posture would provide an idea of change in strategy.

A retaliation only doctrine accepts the fact that India will not initiate a nuclear strike under any circumstances. By extension this also means that India would not use or threaten to use nuclear weapons against a non-nuclear weapon state. On the other hand, in order to reduce the risk of an attack on India, it promises in unequivocal terms that sure and certain retaliation will follow such an attack. Assuming that rationality operates, as indeed it is expected to in any deterrence equation, this prospect which would lead to unacceptable punishment to the aggressor, is expected to function as a deterrent.

Third, as a defensive doctrine of no-first-use and limiting the use of nuclear weapons to retaliation only, India's nuclear doctrine is only a precaution against use of nuclear weapons against it and is in total harmony with the UN Charter. Article 51 of the Charter under Chapter VII clearly endorses *"the inherent right of individual or collective self-defence if an armed attack occurs* against a Member of the United Nations" (emphasis added). Requirements of self-defence against a nuclear weapon threat can either be met through global nuclear disarmament (which would eliminate the means of threat generation) as required by Article VI of the NPT and numerous other international commitments, or achieved through possession of capabilities that could deter the use of nuclear weapons (there being no other credible means of defence against nuclear weapons). In fact, UN Security Council Resolution 1172 is a gross violation of the UN Charter since it seeks to deny India the right enshrined in the Charter and eliminate capabilities that the permanent members of the Security Council themselves possess and jealously guard by disregarding their treaty obligations of nuclear disarmament.

Fourth, the doctrine is based on the concept that the nuclear policy, strategy and posture would be based on the *minimalist principle.* The emphasis on "minimum" deterrence clearly defines this principle in relation to the capability sought, the size of the arsenal, the costs involved, the level of retaliation required and the nuclear posture in peace-time and in times of crisis and active threat. And, hence, the emphasis on "minimum." It is unfortunate that in spite of repeated emphasis on minimum deterrence, even simple statements of intent to respond to a nuclear attack with punitive retaliation are interpreted as "massive retaliation" by knowledgeable experts and reputed organisations well versed in the lexicon and concepts of nuclear strategy.[3]

Fifth, there is one problem that needs to be considered. The final doctrine approved by the government stipulates that India would reconsider the doctrine and could use nuclear weapons in case of a major use of chemical and/or biological weapons. This dilutes the no-first-use doctrine significantly. But what is even more problematic is that this modification to the NSAB doctrine can hardly be treated as credible, and, hence, raises the fundamental question: have our policy-makers understood nuclear deterrence? Both the potential adversaries are parties to the international treaties to abolish such weapons. If there is any doubt about their compliance with the treaty terms, it is possible to ask for inspections, and those for chemical weapons can be extremely intrusive. Any large stocks of such weapons, therefore, would be risky and problematic to hide till the time they are to be used. It is also questionable whether a country would use such weapons since the international norms and opprobrium for such use would be extremely high, and the military advantage marginal, if at all. But assuming that such states retain a clandestine capability to use chemical/biological weapons, the question of proportionality would need to be given serious consideration. And if we still go ahead with the use of nuclear weapons, surely retaliation by the adversary with similar weapons is to be expected. What the strategy implies is that we would have initiated steps for rapid escalation to a nuclear exchange with predictable effects.

On the other hand, the most plausible scenario where chemical/biological weapons might be used, especially when India promises to consider retaliation with nuclear weapons, would be by terrorist groups. In such situations, it is not clear who the nuclear weapons would be used against in retaliation to a terrorist strike with chemical/biological weapons? And since the use of nuclear weapons would be extremely problematic in view of the basic problem of identifying an appropriate, justifiable target and the obvious risk of escalation to an apocalyptical nuclear exchange, the flaws in such a doctrine become transparent. Even if the government considered the threat of use of chemical/biological weapons as credible and serious, a doctrine based on ambiguity might serve the needs of deterrence and catering to its failure much better.

And last, but not the least, the doctrine emphasises that global, verifiable and non-discriminatory *nuclear disarmament is a national security objective.* This is to be expected from a country that has pursued the goal of complete abolition of nuclear weapons for decades in order to remove its own security dilemma as well as ensure greater prospects for international peace and security. In fact, the doctrine clearly states that the absence of nuclear disarmament has created the need for India to acquire its nuclear deterrent. It also makes clear that India's nuclearisation has emerged out of the failure of the international community to institute disarmament half a century after nuclear weapons came into use and more than three decades after the five nuclear weapon states were bound by treaty obligations to do so. It is obvious that if universal non-discriminatory nuclear disarmament were to be instituted, India would be more than willing to give up its own weapons.

On the other hand, there is a view in the country that disarmament was all right as a diplomatic ploy, but if it were to actually take place, India would be much worse off. This is based on flawed logic. If there were no nuclear weapon impinging on our security environment, India would be intrinsically superior to its major adversaries. Even if that superiority is questioned by some, the answer is clear that unambiguous conventional military superiority can be built up and maintained at much lower costs and complications than possessing nuclear weapons. It is obvious that India would remain conventionally superior to Pakistan and, in fact, can achieve even a higher level of superiority at marginally higher costs. While conventional wisdom would have us believe that China has a much larger military force, and given its economic growth of the past two decades, and the massive military modernisation in progress, China would be superior in conventional terms and, hence, our security would remain at a disadvantage if nuclear weapons were taken out of the calculus. But an objective analysis (independent of a 1962 War syndrome) would reveal that this would not be so. India would remain superior in terms of military capability *deployed on the border* as long as India takes prudent precautions to ensure that. Even after the Golmud-Lhasa railway line has been built, serious logistics constraints would continue to affect Chinese usable capabilities. At the present rate of change, the Chinese Air Force would certainly acquire substantive capabilities for power projection in the coming years and this could alter the air power balance with far reaching impact on the overall military paradigm. But this is much easier rectified at lower costs and with less complexities than maintaining nuclear weapons.

Nuclear Arsenal

The doctrine implies that the nuclear arsenal would be so shaped that it provides for a credible minimum deterrent against a nuclear weapons' threat and use. There has been a view that the doctrine promises an open-ended build-up of the nuclear arsenal by India. The most often cited reason in support of this contention is that the doctrine is similar to that of the superpowers because it seeks to build an arsenal based on a "triad of aircraft, mobile land-based missiles and sea-based assets." But the reason for the superpowers' massive build-up was not because of the principle of triad, but because of the overall nuclear doctrine to deter war itself and conventional as well as other forces equipped with weapons of mass destruction. The reference to the sea-based component of the triad has been interpreted as an indication of a large if not huge arsenal. But it is clear to anyone who examines the issues objectively that with a no-first-use doctrine and strategy, *survivability* is the key to the success of the deterrence strategy. And survivability is what gets enhanced with a triad strategy. In fact, a sea-based (particularly a nuclear-powered submarine-based) arsenal would be more survivable even with a smaller arsenal. On the other hand, it would take quite some time for India to possess an operational sea-based (essentially submarine-launched nuclear ballistic missiles) deterrent. In the foreseeable future, the operational force is likely to be confined to the other two components of the triad.

But even there, land-based mobile ballistic missiles with the requisite survivability which can reach the required ranges with the type of accuracy and reliability so essential to credible deterrence would take a long time before they are available as a reliable operational system. Our technology demonstrator Agni missiles, with only two out of three tests succeeding, achieved a success rate of just about 66 per cent, a level of assurance far below any desirable, leave alone necessary, specification. Subsequent tests have shown better results. But even then very few tests have actually been conducted with a solid-fuel ballistic missile capable of ranges beyond 700 km. On the other hand, what we would need are ballistic missiles of ranges up to around 5,000 km to ensure that we have flexibility in selecting the launch sites anywhere in the country so as to enhance their survivability. This would imply that we might acquire reasonably reliable missile systems of ranges of around 3,000 km by about 2010 and ranges of around 5,000 km only after 2015 or so. As far as the submarine-launched ballistic missile capability is concerned, pragmatically we may except another 15-20 years to elapse for such capability to be effectively available as a weapon system. It is worth recalling that the Chinese tested the submarine-launched ballistic missile in 1982 and are not believed to have reached a reliable capability.

There have been reports off and on about India acquiring ICBMs. There are people who believe that India must acquire them, and that without them the nuclear strategy is inadequate, if not flawed. Others simply talk about such delivery systems without carefully looking at the definition of, and treating, all long-range missiles as ICBMs. The accepted terminology of ICBMs assumes ranges in excess of 5,000 km (missiles of ranges below that come under the category of IRBM and SRBM) and mostly in the range of 8,000 km or more. The basic point is that the range of the delivery system would depend upon the location of the target country which we expect to deter. ICBMs would be required if, for example, the US, the UK or France are the points of reference of our nuclear strategy. Such an assumption must first pass the test of political rationality which must make a clear case for the reasons why this is to be so, beyond a vague expression of defiance. At the same time, assessment of costs in political, economic and military terms must be objectively assessed in relation to the purpose sought to be achieved.

There is also a need to look at nuclear weapons specifically for use on the battlefield. In spite of a massive amount of literature on nuclear weapons and strategy, the concept of tactical nuclear weapons or even battlefeld weapons remains imprecise. What is clear is that in our circumstances, and in most others in the world, except where nuclear weapons are being used on the territory of some other country (as NATO strategies required), any use of so-called tactical nuclear weapons would inevitably produce a strategic impact. This makes a classification of some nuclear weapons of defined yield or delivery system misleading, with every risk of false assumptions being created about the use and usability of such weapons below what might be seen as a full-scale use.

In specific, the scenario involving an adversary, say Pakistan, using small yield nuclear weapons focussed on Indian forces advancing in its territory, is often mentioned. Any collateral damage would be in Pakistan and, hence, reduce the pressure for India to retaliate with nuclear weapons. India would then be faced with the dilemma of risking a counter-strike on its vital areas like the national capital, if it responds with counter-value strike(s) or limiting it to a counter-force strike in a similar way. This was no doubt specifically the type of scenario considered in the formulation of the NSAB doctrine in 1999. The doctrine unambiguously states that India would retaliate with nuclear weapons in case of a nuclear attack on India or its forces. It leaves the choice of the weight and target of attack to actual strategy-making. It can be argued that it would be better to retaliate with a similar type of weapon against the enemy's military forces in the field rather than risk a full-fledged nuclear exchange. But we need to make a judgement about whether the use of nuclear weapons can actually remain confined to the battlefield or is inevitably going to escalate to a full-scale exchange? The conclusion is inevitable, that any use of nuclear weapons, small or big, counter-force or counter-value, in own territory or that of the adversary, would inevitably lead to a full-scale exchange of nuclear weapons. What the current doctrine promises is that the enemy would have to take into account the assurance that India would retaliate with nuclear weapons aimed at causing "unacceptable" level of punishment of the aggressor who uses nuclear weapons against India or its forces, wherever they may be. Conventional strategy, of course, would have to take into account the possibilities and provocations for the enemy to reach out for nuclear weapons in certain circumstances.

The draft doctrine makes it clear that India will not follow the disastrous route of the first tier nuclear weapon states which built up tens of thousands of nuclear weapons in a mindless arms race which they find difficult to even reverse. On the other hand, questions have been raised about the doctrine's impact on the arsenal, which need to be addressed. Apprehensions have been expressed that this doctrine would set off a **nuclear arms race.** It is indeed surprising that a doctrine that repeatedly affirms the principles of no-first-use and use in retaliation only is considered by some people to trigger an arms race! Any objective examination of the concept would reveal the reverse to be the case. For a moment let us assume that all nuclear weapon states follow the principle of no-first-use enshrined in the draft doctrine. In that case, no country will initiate a nuclear strike. Logically this implies that nuclear weapons would become obsolete and, hence, should trigger an arms race in reverse, reducing arsenals to minimum levels, and under-cut the necessity for deterrence, if the doctrine is implemented honestly. The doctrine promises that India will work for an international treaty banning the first use of nuclear weapons.

Comparative Doctrines

The central question that has existed, and will continue to exist in future, is: why should we adopt a "no-first-use" doctrine? We must face the unthinkable squarely: after all, are we going to wait for nuclear bombs to be dropped on our country before we launch our own?

To get answers to these, and many other questions, we must be clear in our mind: why do we need nuclear weapons? The only rational, sensible answer is that we need them to deter another nuclear weapon state from using them against us—and nothing else. This is based on the reality that nuclear weapons cannot really deter other forms of violence, like terrorism, as even the US realised on September 11, 2001. There are broadly two ways of achieving the goal of nuclear deterrence: either by a policy of using them first, or by a doctrine based on no-first-use. If the aim were to deter the use of conventional military power, then a first use strategy would be necessary. But surely, India has sufficient capabilities (and capacity to enhance them) to successfully deal with any and every conceivable conventional military threat by conventional means at a level of investment of 3 per cent of the GDP (Gross Domestic Product).

But to understand the logic of a no-first-use doctrine, it would be useful to look at the implications of using nuclear weapons first. To start with, the history of the past half century tells us that credibility of nuclear deterrence relies heavily, if not completely, on the ability of a nation to undertake a "second strike," that is, the ability to possess sufficient weapons which would survive the enemy's first nuclear strike to be able to retaliate and cause unacceptable levels of damage and destruction. This is expected to convince the adversary that its use of nuclear weapons would lead to its own assured destruction. It has also been clear for decades that there can be no certainty that the adversary's nuclear arsenal can be completely destroyed in a first strike. This led to a series of historical developments.

The first policy inevitably led to increasing the size of arsenals in a mutually debilitating arms race, which at one stage resulted in the existence of nearly 66,000 nuclear warheads mostly owned by two countries but deployed globally. This experience is at the root of concerns of a nuclear arms race in our region. But China did not enter that race because it did not opt for a first strike option and yet deterred both superpowers. Surely, it would not serve our interest to start building an open-ended nuclear arsenal.

The first strike policy also requires nuclear weapons to be kept on hair-trigger alert to launch them within minutes, if not seconds, exposing them to an enemy strike. The race is then to launch them before the adversary launches its own weapons. Preemption becomes a key element in such a nuclear strategy. Given the huge arsenals, the extremely short time available, and the need to hit the other side first, this requires an elaborate command and control system, and its attendant complexities, costs and technologies, which currently exist only with Washington and Russia. Such complexities also inevitably increase the risk of accidents, unauthorised use, less than adequate control (as in the case of submarines), or use due to miscalculation.

But the most important lesson is that if a second-strike is the acid test of nuclear deterrence, then why not rely on retaliation only? Assured retaliation amounts to a second strike although no first strike would have been launched by us. But like the second strike strategy, it requires the ability to launch sufficient nuclear weapons after the enemy's nuclear strike. Even the simplest logic would tell us that survivability is the key to credible nuclear deterrence, and not the first strike policy *per se*. And this must be capable of assured retaliation that would cause what is euphemistically called "unacceptable punishment" to the aggressor. The no-first-use doctrine is the only way to ensure that they remain in a defensive role, to be used only for counter-attack.

We must also objectively assess what would be the scenarios where India would need to use nuclear weapons first, warranting a first-use strategy and its attendant costs? A significant military attack, if not a decisive military defeat, or surprise breakthrough by the enemy with conventional forces are two possible theoretical contingencies. But these could happen only if we allow our conventional military capability to go down to unacceptably dangerous levels. By any logic, therefore, a no-first-use doctrine is not only the best option for us, but also keeps the nuclear genie under better control.

There is no doubt that India's nuclear doctrine poses a serious challenge to the prevailing doctrines of offensive orientation and the first strike strategic doctrine of the US/NATO and Russia. China has been demanding a treaty on no-first-use among the weapon states. The Chinese and Indian doctrines now indicate a counter view to the traditional aggressive doctrines of other nuclear weapon states that visualise use of nuclear weapons against non-nuclear threats. It is not surprising, therefore, that the US responded to the draft doctrine by saying that India is "moving in the wrong direction."[4] But since the State Department spokesman acknowledged that the US had yet to examine the document, this appears to have been a reflex action to a doctrine which reflected thinking in total contrast to the US doctrine and belief system.

- The credibility of nuclear deterrence would remain unaltered with the NFU since assured retaliation would be similar to a "second strike" capability as the defining element of credible deterrence.
- The NFU strategy would avoid most of the risks associated with nuclear weapons like:
  - Accidental launch.
  - Miscalculation.
  - Preemption risks with hair-trigger alerts.
  - Escalation to mutually assured destruction.

- A graduated response in the escalatory process is possible. The onus of escalation lies with the adversary and provides diplomatic advantage in dealing with the crisis.
- It would drastically reduce, if not completely eliminate, the coercive role of nuclear weapons since threat of use is incompatible with the NFU strategy.
- It reduces the salience of nuclear weapons in security and foreign policy.
- NFU strategy constitutes an important forward step on the path to the abolition of nuclear weapons, especially if accompanied by other measures like banning the use and threat of use of nuclear weapons (similar to the 1925 Geneva Convention on chemical weapons).

The no-first-use strategy, by definition, establishes a strategy of non-use of nuclear weapons against non-nuclear weapon states, however they may be defined. This may not be seen as an adequate security guarantee by itself. But as China has proposed, a No-First-Use Treaty would provide greater assurance and security than the existing promises which carry many caveats.

Command and Control Issues

There are accepted norms of command and control systems for military formations. Military power itself is an instrument for achieving political objectives and, hence, constitutes a political instrument of the state. At the other level, it has to be employed as a military instrument. Similarly, nuclear weapons essentially exist as a political tool which must also, in the extreme case, be employed as a military weapon, although with little military utility as such. To maximise the political utility of nuclear weapons, therefore, the following principles must be kept in mind:

(a) First, firm political control over the nuclear arsenal. I will revert to that in a moment.

(b) Second, a highly efficient and responsive military command and control system. With that I mean that decision-making must remain at the highest political level with the advice of the military and other experts, but essentially the command and control system as such must remain with military force for a series of reasons.

(c) The third principle is that there must be minimum disruption of well-tested existing organisations and procedures. Our goal should be to see where they need to take on additional responsibilities, with minimum modification. It would be a gross mistake of truly monumental dimensions to attempt an organisational revolution to manage a multiple set of things—fundamentally new type of weapons, a new set of command and control systems and a new set of procedures to handle these. Any major experimenting or violation of basic principles of controlling military systems will lead to serious and irreversible problems for many years, perhaps decades, to come. Therefore, we must build on existing systems and procedures wherever they exist and do so in an evolutionary form, review it constantly as we go along and then at a future stage, if we still need to introduce significant changes, we should do so in an evolutionary, planned manner. A conservative approach will be well worth the effort. We are entering a stage where not enough thought has been given to the details of the circumstances and the way in which nuclear weapons are to be used. The military in particular has been kept out of most of the decision-making and examination of the implications in this process. The case of chemical weapons is symptomatic. Therefore, it is essential that we move in to the nuclear weapons command and control through an incremental, evolutionary process built on existing capabilities and past experiences.

(d) The fourth principle is that costs must be minimised, otherwise we are likely to land up with an expensive system which may or may not work. Experimenting or rushing into something here is the last thing we should be thinking of. If that is so, do we have any models that exist at least partially, in which the military at least has handled similar tasks in the past? I would like to draw attention to the command and control of "Chiefs of Staff Targets" which have been effectively managed for more than four decades, in peace and war. The three Services under the sanction and direction of their chiefs have evolved the framework and procedures which authorise their staff to work out the targetting policy for a set of targets considered of central and strategic value, as the strategic targets to be released under the combined authority of the three Chiefs of Staff. Detailed targetting and planning for this purpose has been maintained by the Indian Air Force over the decades, updating them as and when required by the joint planning process of the Chiefs of Staff Committee. The planned air effort of the air force, obviously, accords requisite priority to commitment for Chiefs of Staff Targets. The concept, in principle at least, is no different than targetting with nuclear weapons. Targetting with nuclear weapons needs additional factors to be considered, and greater involvement of the three Services will be necessary. But the basic requirements and parameters of management are not very different, either in terms of organisational needs or management techniques. Secondly, the Indian Air Force has been undertaking, like those in other countries, strategic reconnaissance missions over the decades. The way command and control has been managed, and the lessons derived from decades of highly successful operations would be extremely relevant to the issues related to command and control of the nuclear arsenal.

(e) The fifth principle is that the bulk of the work related to effective management of the nuclear arsenal has to be done during peace-time. This involves scenario building and assessing the other country's postures, likely trends, changes that may take place, signalling of intent, etc. For example, what are the implications when the Chinese say during a crisis situation that "Los Angeles could burn?" How seriously should such pronouncements be taken? Do they constitute a firm form of escalation moving towards use of nuclear weapons or not?

(f) Last, but not the least, is the issue of assessment of what it will take to deter the other side under different sets of circumstances. This will remain a matter of judgement, but one derived from extensive and deep study of the cultural, socio-political and strategic factors affecting the likely response of a country to the threat and use of nuclear weapons. I am sure many more aspects and principles can be added to what the command and control system should be expected to do, and what it can do in the framework of a time horizon.

Having considered this and a few other factors, it is obvious that at the political level, command and control of nuclear weapons, in the ultimate analysis, must be exercised by the prime minister of India. Given our political constitutional system, it cannot be anybody else. This can change only if the constitutional basis of our governance changes. A concurrent issue is that there is a need for a consultative group or decision-making group to assist the prime minister at that level of political leadership. This has two aspects. One relates to the chain of command to cater for a contingency if the prime minister is not available or is incapacitated. The Constitution lays down the succession process very clearly. But the process may be inadequate since it did not visualise the need to plan a response with nuclear weapons. So we have to evolve a suitable system of national command authority. Whatever system is evolved, the prime minister of India will have the prerogative to make changes to it. What is required is that each prime minister, at any one time, should have a well-defined system of national command authority for command and control of nuclear weapons. Which one it is, is not the point. It may also not be possible to formally declare the system, nor is it necessary, as long the operating formations that have to execute the orders are clear about the chain of command. In fact, there are strong reasons for not publicising the chain of command for obvious reasons besides the potential for domestic political difficulties of the type experienced in the past. The prime minister, who selects his Cabinet ministers, will always have the prerogative to make even frequent changes to the chain of command, but he will have to ensure that the new person(s) concerned are briefed thoroughly and they understand the role fully.

At the same time, a decision-making-cum-planning body at the political-strategic level is needed which will then authorise the planning of targetting policy and the force development and deployment policies. If the NSC functions, then the Cabinet Committee would be the obvious organisation. Even in that case, there is a need to immediately resurrect the Defence Committee of the Cabinet (DCC) with the three chiefs in attendance, on permanent basis, not just on invitation, which should start taking the planning process forward. Presence in attendance implies that they will not participate in the deliberations of the Cabinet, but will be available to answer queries and provide explanations where needed.

It may be recalled that when the higher defence organisation was reorganised in 1947, it was based on the experience of the UK and the USA (the latter being a nuclear weapon power) during World War II. In fact, Lord Ismay, who planned the post-War higher defence organisation for the UK and the USA was also the person who planned the organisation for independent India. This was based on corporate decision-making at each level with a Cabinet Committee at the apex, a Defence Minister's Committee, the Chiefs of Staff Committee (COSC), Joint Planning Committee, a Joint Intelligence Board and another dozen committees of the government to take decisions concerning various aspects of defence. Interestingly, the Joint Planning Committee was to be in permanent session with a permanent staff "in particular, to prepare plans for the joint employment of the three Armed Forces." Unfortunately, the permanent staff was never set up. In 1986, the government established a joint Defence Planning Staff of the Chiefs of Staff Committee. However, its relationship with the Joint Planning Committee was never clear. The result was that the operations branch of the individual Services kept operating nearly autonomously. This, in turn, created numerous incidents that have tended to increase the gap of natural understanding of roles and missions, especially between the army and the air force. Over the decades, the effort of the army and the navy then has been to try and bring as much of the air power as possible under their own control. But classic cases of failure of joint planning naturally arose out of the failure to establish the requisite structures for this purpose which in turn also adversely affected the development of a culture of jointness that is so much talked about but rarely practised these days.

There is a need to address the question: where would the inputs to the DCC or the Cabinet Committee on National Security come from? These inputs will be a mixture of the military elements and political-diplomatic implications of decisions. Thus, an increasing amount of military component gets introduced into the command and control system as we move down the ladder of national command and control of the nuclear arsenal. The logical locus of such component of the management system is the Ministry of Defence. Other ministries and departments, like the Finance Ministry have an important role to play in the force planning though their role in force employment would perforce be limited. In my view, the Defence Minister's Committee (DMC) should be resurrected (with some modifications) at an early date with the clearly defined role of decision-making about strategic planning and force structure planning, including that for nuclear weapons. The DMC would naturally be chaired by the defence minister and have as its members, the minister for state for defence, the three defence Service chiefs, the defence secretary, foreign secretary, secretary DAE (Department of Atomic Energy), home secretary, SA (special assistant) to the defence minister, chairman JIC (Joint Intelligence Committee), and the finance secretary. The DMC should be logically served by a professional strategic planning staff under the defence minister which should be able to situate the nuclear strategy and policy within a broader strategic planning for defence.

Below the ministerial level, the requirement shifts to more military-specific planning and direction—the nuts and bolts, so to say. This is really what should be treated as a task of what may be termed as the Nuclear Planning Group which would work out the detailed targetting policy. This is the sort of function that the Indian Air Force has been undertaking for decades in respect of COSC targets. The Nuclear Planning Group should preferably be headed by the defence minister himself with the three Chiefs of Staff, SA to the defence minister, and the defence secretary as members. The Defence Planning Staff is really now in a position to perform the role of an integrated staff and provide staff support for the Nuclear Planning Group that will go into detailed planning in respect of target policy, dispersal policy, deception policy, survivability policy, communication policy, etc.

Operational Command and Control

That takes us down to a different level, that is, operational command and control of the nuclear forces. In this case, the authority for exercising the overall operational command and control over the nuclear arsenal, and certainly the control aspect should rest with the COSC. If and when we have a permanent chairman Chiefs of Staff or a Chief of Defence Staff, then obviously he will become the spokesman for the COSC, but even then a Defence Planning Staff will be needed which should operate through the COSC.

Below the planning and management level what we need is an operational force with requisite command and control to suit the doctrine and force structure. For quite some time, the primary nuclear delivery system will remain the combat aircraft. The overall Indian Air Force (IAF) infrastructure provides a sound building base for implementing the strategy to operate the nuclear strike component from widely dispersed sites and maintain deception and dispersal strategies. Conceptually, the task at the operational level is the same as that performed by the IAF day after day. Strike aircraft of today are capable of performing dual-role (conventional as well as nuclear weapon delivery) tasks. The optimum approach would be to reassign some elements of the IAF strike aircraft assets for the nuclear delivery role. In order to optimise the command and control capabilities, it would be useful to create a Strategic Air Command through reorganising the existing assets. Very little change will be needed to create a Strategic Air Command which would have operational command and control of nuclear delivery assets. Such an organisation would be ideally capable of absorbing surface-to-surface ballistic missiles into its structure as and when they are ready for induction—a process that is bound to be spread out over time—and be able to keep taking in and absorbing surface-to-surface missile systems as we go along and they are inducted into service. The reasons for this approach are fairly simple. This will have the main strike aircraft component that will also obviously have a dual role. Perhaps we are talking about PGMs being delivered, possibly to take out even a nuclear facility or elements of the nuclear arsenal of the other side. Where do we draw the line between the strategic strike with conventional weapons, PGMs and strategic strike with nuclear weapons, especially of any different yield?

The surface-to-surface missile, especially land-based, in terms of operational utilisation, in terms of its management, in terms of its command and control, is no different from the strike aircraft. We must make sure that we do not have strike aircraft heading on their own under a different command and authority, and surface-to-surface missiles under a different one, and attempt coordination at that stage. That could spell disaster. There has to be unity of command of strategic strike capabilities and, hence, a single operational command to employ capabilities, including nuclear weapons. This will obviously include suitable electronic warfare systems, requisite strategic reconnaissance systems, a suitable space-based and ground-based electronic early warning system, and if and when we get them, anti-ballistic missile systems. Should this remain an Air Force Command totally? I don't think so. I think it should have on its staff officers from the other Services to be the primary advisors to the commander of (conventional and nuclear) strategic forces. What are the larger implications of this beyond turf battles and loyalties?

The third element concerns the assumption that seems to have gathered strength over time that the nuclear warheads would or should be held in the jurisdiction of the scientists. The warhead making process, its maintenance, its overhaul, its efficiency and everything that goes into it, certainly is a scientific job, to be managed within the scientific establishment. Also, for a different reason, which is more related to nuclear arms control and nuclear diplomacy, I recommend that we establish a clearly demarcated military facility within the Department of Atomic Energy so that warhead construction and maintenance can be linked to the military side without interfering with the civil programme for peaceful purposes. I firmly believe, and it also needs to be debated in the years to come, that the defence forces must hold the nuclear warheads. If nuclear scientists are required for field maintenance and handling, they should be coopted from the atomic energy establishment. But they will have to be brought under the Army, Navy or Air Force Act.

To handle safety and the management and maintenance of the warheads at the end point, the main work should continue to be undertaken by the Department of Atomic Energy. There are two issues to be considered. Firstly, who would control the warheads? Secondly, this will also decide who releases them to those who control the delivery systems. In my view, the warheads should be controlled by the Indian Army. The Indian Army is the most suitable agency really capable of providing the type of security and physical safety to the warheads that will be required in highly dispersed conditions across the country, with many of them being shifted in a variety of ways. The Indian Army itself will have to create an elite type of force for such a purpose for it is quite likely that for every one actual warhead, we may require a number of dummy warheads which will still have to be protected the same way. That is the only way we can enhance the survivability. There are lots of lessons we can draw from the experiences of the air forces of the world, not just our air force, on how they manage their deception and duplication plans. It is useful, for example, to see how Iraq managed the deception with dummy missiles and the movement of missiles all over the place. It is an essential element. I think the army has the basic ability and organisational capabilities to handle it. There would be a parallel chain of command where this links up with the HQ of the Strategic Command to be able to ensure that the release of nuclear warheads is carried out only when there is an agreement among the senior air force man, the senior army man and the senior naval man there. This would build in security locks in terms of utilisation of warheads in a contingency.

Communications are going to be the crux of the command and control systems and they will also determine how the system will operate in an emergency after absorbing a first strike. The starting point is to accept the philosophy and doctrine that the nuclear arsenal will be widely dispersed, mobile, moving around at frequent intervals, so that at no stage can the adversary have a clear idea or possibility that he can actually decapitate the nuclear capabilities in a preemptive strike. So we need to make sure that the first strike by an adversary does not degrade our capabilities to an extent that we cannot impose a counter punishment that is unacceptable to him. Deterrence requires that the adversary should be sure that there would be a substantive surviving nuclear arsenal that will hit back within a very short period. This requires gearing up of communications systems. We may have to seriously think of setting up an integrated Defence Communication Command (or agency) under the COSC where obviously the Indian Army would be the major partner. We must ensure compatible and inter-operable communication systems and frequencies.

One last word. A lot of parameters would be altered if tactical nuclear weapons were introduced into the equation since this will alter the requirements of command and control very fundamentally. So that is one decision that must be made early on and then a clear posture, publicly taken on that issue, must be adopted. We must not even have any suggestion that we will have tactical nuclear weapons or that we will in any way engage in the use of nuclear weapons in a battlefield situation. If we have nuclear warheads of sub-kiloton range, they are perfectly usable on very specific, specialised target systems, but not in terms of tactical or battlefield use. Nuclear deterrence must rest firmly on assured retaliation in strategic terms, finally but clearly linked to a response that will cause unacceptable damage to the potential aggressor even with the surviving arsenal, after absorbing a first strike.

What is even more critical is that as long as India adopts a doctrine and strategy of no-first-use, survivability of the nuclear arsenal would be the key to the efficacy of the credible minimum deterrent. It is also clear that the nuclear arsenal would be the primary target for an adversary who must aim to take out the arsenal if it is to preempt a retaliatory strike and its punitive effects. It would, therefore, be vital to ensure dispersal and deception of assets to enhance survivability. If we are to designate the nuclear strike force from within the present size of strike forces and place it under a separate operational command authority, we would be creating conditions for presenting a smaller target system for the adversary's preemptive decapitation strike and, thus, making its task easier. Under these circumstances, we would not only be going in for a significant quantity (and proportion) of idle (nuclear armed) strike assets which would be required only for nuclear retaliation, but also risk erosion of survivability of the nuclear strike force.

Strategic Stability

It also needs to be recognised that the risks of a nuclear threat and possible use increase with the prospects of an inter-state war where one or both sides possess nuclear weapons. A conflict situation, therefore, lowers the nuclear threshold by increasing the risk of possible use. India's interests are better served if there is no nuclear threat. This is why abolition of nuclear weapons is perceived as a national security objective. In the absence of nuclear disarmament, there is a need to keep the nuclear threshold as high as possible so that the nuclear threat factor does not enter into India's operational security situation. India's interests would be prevention, and avoidance of war would naturally raise the nuclear threshold. It would also be necessary to keep conventional military capability sufficiently high to provide a credible effective deterrent against conventional war. This would help to raise the nuclear threshold. It can, of course, be argued that if India maintains a higher level of conventional military capability than before, this would increase the threat perceptions in Pakistan which may then tend to lean even more toward early reliance on nuclear weapons. This only enhances the importance of the credibility of the Indian nuclear deterrence.

On the other hand, there is no objective reason for threat perceptions in Pakistan to increase since its need to reach out for the nuclear button could rationally arise only in case of an Indian invasion succeeding, which it can no longer hold with its conventional forces. But Pakistani leaders will have to face up to the truth that all wars so far since 1947, including the latest one in the Ladakh sector of Jammu & Kashmir state, were initiated by Pakistan; and all that Pakistan has to do, therefore, is not start another war. But by the same logic, the reverse would be true in the case of China which in any case is modernising and upgrading its military power in a significant way.

Notes

1. *Arms Control and Disarmament,* White Paper, Information Office of the People's Republic of China, Beijing, November 1995.
2. See Ashley Tellis, *India's Emerging Nuclear Posture* (Santa Monica: RAND, 2001), pp. 211-226.
3. Introduction to "India's Nuclear Doctrine," August 18, 1999, in *BASIC Publications,* http://www.basicint.org/indiadoct.htm dated September 3, 1999, distributed as Background Document at 49th Pugwash Annual Conference, September 7-13, 1999, Rustenburg, South Africa.
4. Aziz Haniffa, "Clinton Writes to Vajpayee, US Calls N-Doctrine a Wrong Step," *India Abroad News Service,* August 19, 1999.

12

Disarmament in India's Nuclear Strategy

Manpreet Sethi

India is now a state with nuclear weapons, irrespective of whether the Western nuclear powers recognise it as such or not, shackled as they are by the narrow confines of the NPT definition of an NWS.[1] In May 1998, after India had conducted the five nuclear tests, the Indian prime minister proclaimed the nuclear weapon status of India. However, significantly enough, in the same breath he reiterated the country's commitment to nuclear disarmament. Subsequently, the draft nuclear doctrine prepared by the National Security Advisory Board and put forth in August 1999, also included nuclear disarmament as a cherished national objective for India. This is a statement of immense import since a nuclear doctrine is meant to operationalise a nuclear weapons development and deployment strategy. No nuclear doctrine of any NWS, except that of India, upholds nuclear disarmament as its ultimate objective. What, then, does this signify?

Five years and a few months after the dramatic change in India's nuclear status, it may be useful to objectively analyse the country's stake in the pursuit and attainment of the goal of nuclear disarmament. Before India conducted the tests, there were many who had dismissed India's initiatives towards the elimination of nuclear weapons as a case of sour grapes. It was said that since India did not possess nuclear weapons, it aspired for their elimination so that no one would have the advantage. After having demonstrated its nuclear weapons capability and having decided to weaponise, there are many who argue that the country must now give up its moralist stance on nuclear disarmament and concentrate on building an effective and reliable deterrent.

This chapter, however, seeks to argue that for India, its struggle for universal nuclear disarmament goes beyond the rationale spelt out by these two arguments. The grapes were not sour when India did not have nuclear weapons. It was just that during those 25 years of restraint and earlier, India believed that its national security was better served without nuclear weapons. But a steady deterioration in the regional security environment, owing to a growing China-Pakistan nuclear nexus, and a total lack of response to Indian and other initiatives towards nuclear disarmament made it necessary to reassess the country's security requirements and move towards the acquisition of these WMD. This was well articulated by Prime Minister Vajpayee when he said, "All the governments in New Delhi since independence have steadfastly maintained that our national security as well as global security is guaranteed only in a nuclear weapon-free world. India backed up this conviction with a large number of initiatives in the United Nations and other forums...Sadly our initiatives were disregarded by the NWS...Yet, India waited 50 years before responding."

In the wake of vertical and horizontal proliferation in the neighbourhood, India was left with little choice but to go down the nuclear path itself since the only effective deterrent for nuclear weapons are nuclear weapons themselves. But, if India's security requirements can be catered for through a universally acceptable and verifiable disarmament regime, India need not hesitate to disarm along with the others. In fact, to do so would be in its larger interests and it is for this reason that it makes sense to continue efforts for the realisation of an NWFW.

This chapter is broadly divided into three sections. The first one historically traces the major Indian initiatives on nuclear disarmament. The second section examines the factors responsible for bringing about a change in mood within India on the issue of nuclear disarmament. The final section then argues why it is still in India's interests, despite the change in mood and the country's status as an NWS, to strive for disarmament.

Indian Initiatives: A Historical Assessment

India's relationship with nuclear disarmament goes back a long way. Ever since independence, Indian governments have made use of every opportunity and forum to press upon the international community to treat the subject of nuclear disarmament with the kind of seriousness and urgency that it deserves. In fact, even before India joined the comity of nations as a sovereign country, Jawaharlal Nehru, who was later to become the country's first prime minister, had said in 1940:

> Both because of our adherence to the principle of non-violence and from practical considerations arising from our understanding of world events, we believe that complete disarmament of all nation states should be aimed at, as in fact an urgent necessity, if the world is not to be reduced to barbarism.[2]

In 1945, when the *Little Boy* and *Fat Man* were first dropped on Hiroshima and Nagasaki, Mahatma Gandhi realised the impact of this new weapon, though he chose not to speak against it, fearing that the British rulers could use it to deny independence to India. He later said that the bomb would bring moral devastation on those who developed and used it. He vowed to fight for outlawing the nuclear weapons. While he could not personally conduct the crusade, every Indian government has upheld nuclear disarmament as a moral imperative since the very nature of destruction wrought by nuclear weapons negates the very spirit of humanity. Even when India was not in possession of the lethal weapon, it did not consider an NWFW as an international abstraction, but as a necessity in whose attainment every nation, big or small, with or without nuclear weapons, had an equal stake.

India's very first articulation on nuclear disarmament after its emergence as a modern nation state provided a unique conceptualisation to the term. Indian leaders upheld it not only as a political necessity for ensuring security and peace, but also as a moral imperative. In order to translate the concept into reality, India tried to make use of every opportunity, within the UN and outside it, through appeals, proposals and actions plans, to get the international community to pursue it consensually.

As early as in 1948, at the time of the consideration of the report of the newly created United Nations Atomic Energy Commission, India proposed limiting the use of atomic energy to peaceful purposes, and the elimination of nuclear weapons from national arsenals. Two years later, and despite its preoccupation with several urgent tasks of nation building, India used its voice in the UN to draw world attention to the fact that the growing arms race was not only a threat to international peace and security but also a drain on human and economic resources that could be better utilised for other developmental activities. To rectify this imbalance, it suggested the creation of a UN Peace Fund formed through a progressive reduction of arms, and directing the amount thus released towards development.

In 1953, speaking before the First Committee of the UNGA, then

Foreign Minister V.K. Krishna Menon had said, "There is only one thing to do with atomic weapons and that is to do away with them". In 1954, and soon after the USA first tested hydrogen weapons in the Marshall Islands, Prime Minister Nehru called for an immediate halt to all forms of nuclear testing and this eventually formed the basis for the conclusion of a PTBT.

In 1955, Krishna Menon once again called for a discontinuance of testing, for a new determination to reach an agreement on reductions, and for an absolute prohibition of the use of atomic weapons. One year later, India presented a memorandum to the Disarmament Commission suggesting certain concrete steps. These included a cessation of experimental nuclear explosions, dismantling of nuclear weapons in order to reverse the arms race, declarations by countries not to manufacture any more nuclear weapons, and prohibition of export or transfer of nuclear weapons to other countries by the then declared NWS. Unfortunately, none of these measures received the attention that they deserved. In 1957, the Indian delegation to the 12th session of the UNGA put forth a draft resolution appealing to the "states concerned to agree without delay to suspend tests of nuclear and thermonuclear weapons." The resolution, however, was defeated by a vote of 34 to 24 with 24 abstentions.[3]

At an anti-nuclear convention held in 1962 in New Delhi, the then Indian president said, "The non-aligned and neutral people are as much involved in this as those who are knowingly engaged in the criminal conspiracy of creating weapons and conditions that would spell their own annihilation no less than that of others."[4] In 1964, India took the initiative to place the item "Non-Proliferation of Nuclear Weapons" on the UN agenda and thereby sowed the first seed of the NPT. It had then called for the treaty to be based on three principles:

(i) Absence of any loopholes that might permit the NWS and NNWS to proliferate weapons either directly or indirectly.

(ii) Establishment of an acceptable balance of mutual responsibilities and obligations of both the NWS and NNWs.

(iii) Marking a step towards general and complete disarmament.

It is a different matter that the NPT that finally took shape was devoid of these principles and has, hence, not been accepted by India to date.

However, despite its disillusionment with the NPT, India continued its efforts into the 1970s. At the UNGA's First Special Session on Disarmament in June 1978, Prime Minister Morarji Desai called for outlawing the use of nuclear technology for military purposes, including its research and development. Four years later, at the second session, the Indian foreign minister proposed a 5-point programme in this regard. It sought the negotiation of a binding convention on the non-use of nuclear weapons, an immediate suspension of all nuclear tests, a freeze on nuclear weapons, attainment of general and complete disarmament within an agreed time-frame and the UN taking the lead in educating the public on the dangers of nuclear war, the harmful effects of an arms race and the positive aspects of disarmament and its link with development.

At another level, in 1984, India launched the 6-nation 5-continent Peace Initiative along with Argentina, Greece, Mexico, Sweden and Tanzania. This measure came at a time when the dialogue between the nuclear powers had collapsed into accusations of ill faith and mutual recrimination. In 1986, during President Gorbachev's visit to India, Prime Minister Rajiv Gandhi used the occasion to once again send out a forceful plea for disarmament. In a joint statement, the Delhi Declaration, both heads of government "on behalf of more than one billion men, women and children..." called for urgent action that would lead to a "world free of weapons of mass destruction, a world without war."[5] In 1988, at the third Special Session On Disarmament (SSOD III), the Rajiv Gandhi Action Plan was unveiled that sought a binding commitment by all nations to a phased elimination of nuclear weapons by 2010; the establishment of a comprehensive global security system under the aegis of the UN; and, simultaneous negotiations on a series of integrally related measures such as cessation of all production of weapon grade fissionable material, a moratorium on testing of nuclear weapons to set the stage for the CTBT, etc.

As is evident, the Rajiv Gandhi Action Plan for Ushering in a Nuclear Weapon-Free and Non-Violent World Order was a well-considered and comprehensive one that identified specific activities that could be realistically carried out in different stages. While the plan found praise from amongst the NNWS, and even some of the NWS, in the absence of sincere politicl will, and the resolute determination of the NWS to cling on to their nuclear arsenals, there was no concrete movement towards nuclear disarmament during the Cold War years.

Post-Cold War, India found its clout on the issue of nuclear disarmament somewhat diminishing with the fall in relevance of the NAM. From the NAM and Soviet Union, India had derived both numerical and ideological strength for its resolutions at the UN. But even as this fell, in contrast, US dominance grew and best came to light when in 1995 it was able to bludgeon several NAM nations party to the NPT into legitimising nuclear weapons through granting an indefinite and unconditional extension to the NPT. Since then, Indian efforts at promoting nuclear disarmament through the UN have largely been reduced to tabling resolutions before the UNGA, support for which has been steadily dwindling. The CD too remains stalemated, at least for the last four years over two issues: the NWS refusing to give in on the demand of the NNWS to establish an ad hoc committee on nuclear disarmament; and the US-China clash of interests on the issue of PAROS.

India and Nuclear Disarmament: A Change in Mood

Given the cold response meted out to a large number of Indian initiatives and a simultaneous increase in the salience of nuclear weapons within the nuclear strategies of the NWS, it is not surprising that, within India too, there should be many who question the wisdom of India's pursuit of this objective. Especially after going nuclear, it is felt that the country must now concentrate on building a credible nuclear deterrent rather than wasting its energies and time on articulating the merits of nuclear disarmament. The position of those who hold this view is based on three factors:

- Spread of cynicism due to the lack of seriousness on the part of the NWS. Rather, a reinforcement of the nuclear component in their military doctrines and a continued reliance on nuclear deterrence have only added to the cynicism. These developments stand out in sharp contrast to the pledge undertaken by the NWS at the time of the NPT extension in 1995 that they would fulfill their obligations under Article VI of the NPT to initiate steps towards nuclear disarmament. At the same time, they also promised not to use nuclear weapons against NNWS parties to the treaty. This promise was repeated by all the five NWS in the joint UN Security Council Resolution 984 (1995), which was unanimously adopted on April 11, 1995.

However, as has become increasingly evident over the last decade, these undertakings were obviously little more than ploys for securing international support for the unconditional extension of the NPT and have since been forgotten. Nuclear deterrence remains the mainstay of the military strategies of the five NWS and is likely to remain so into the "indefinite future."[6] Meanwhile, the US has not hesitated in expanding the core mission of its nuclear arsenal from deterring only nuclear weapons to now using them as a deterrent for chemical and biological weapons also. Russia too has retracted its commitment to not be the first one to use nuclear weapons. Rather, its military doctrine, put forth in January 2000, sanctions the first use of these weapons against even a conventional attack, in case other means at its disposal have been exhausted. The recently disclosed US policy of preemption in case it suspects a country of pursuing a WMD programme retracts US commitment to negative security assurances as extended to the NNWS in 1995. Actions such as these have only served to make other states that do not possess nuclear weapons reconsider their approach to nuclear disarmament and to view it rather cynically in the wake of the ongoing trends.

- The second factor to have impacted adversely upon nuclear disarmament is the growing feeling about its unfeasibility. There is a feeling of inability to tackle issues such as the logistics and cost involved in the dismantling of tens of thousands of weapons, accounting of every gram of fissile material recovered, effective verification mechanisms, etc. Besides, even if the existing nuclear weapons were to be destroyed, the knowledge of bomb making would continue to exist. Therefore, there is widespread scepticism over the feasibility of nuclear disarmament. This sentiment found an echo in a US government document of 1995 in which it was stated, "Since it is impossible to uninvent nuclear weapons or to prevent clandestine manufacture of some numbers of them, nuclear weapons seem destined to be the centerpiece of US strategic deterrence for the foreseeable future."[7]
- Thirdly, there is a takeover of idealism by *realpolitik* in the Indian polity. Idealism has slowly been giving way to greater realism in external relations and foreign policy-making. In this world of *realpolitik*, the ideal of nuclear disarmament has taken

a beating. Of course, it cannot be overlooked that the voices of the realists contesting the official position had existed even as far back as the early 1960s when after India's defeat at the hands of the Chinese, Beijing had demonstrated a nuclear capability. This had given a fillip to the pro-bomb lobby,[8] though its reach had remained limited.

In fact, till 1998, the pro and anti-bomb lobbies had shared room for manoeuvrability as the official Indian position remained one of nuclear ambiguity. The sense of idealism and the principled approach was adequately balanced by the belief that national security considerations demanded the availability of all options. Post-1998, however, idealism has taken a beating as other, more real issues of deteriorating regional security environs have surfaced. Hence, the security dimension of nuclear policy assumes greater sway with a consequent de-emphasis on disarmament. Also, India has seen support for its initiatives dwindling from among the ranks of the non-aligned nations, further reinforcing the perceptions on the impracticality of nuclear disarmament.

India's Stake in Nuclear Disarmament Post-Pokhran II

Notwithstanding the above, there is still need to examine as to why India stands to gain by persevering in its pursuit of nuclear disarmament, however, distant the goal may appear at present. India has long campaigned for it on the grounds of morality and humanitarianism. While this logic shall hold eternally, in the present world of *realpolitik,* it would be worthwhile to argue the case for universal nuclear disarmament on the grounds of India's national security.

Since the decision to opt for nuclear weaponisation was made on the grounds of national security, it is imperative that the decision to disarm must also be based on the same logic. Only if the country's national security is not jeopardised, but is rather enhanced, then and only then, can a persuasive case be made for India's continued pursuit of the objective of nuclear disarmament. And, this is precisely the point that this chapter tries to make in the paragraphs that follow. Of course, it needs to be said that India can accept only *universal* nuclear disarmament since it is only in such an environment that its national security would be enhanced. Unilateral nuclear disarmament is a non-option.

Nuclear Disarmament and National Security

As has been argued earlier in the chapter, India's decision to overtly demonstrate its nuclear weapon making capability was necessitated by a deteriorating regional security environment. The overt and covert presence of nuclear weapons in the immediate neighbourhood made it imperative for India to acquire the same so as to hold forth an effective deterrent. But, if nuclear weapons were to be removed from the neighbourhood, and at the global level, courtesy a universal nuclear disarmament regime, the situation would enhance the country's security. In fact, it is the presence of nuclear weapons, regionally and worldwide, that complicates India's security calculations. And, hence, if they were to be removed from the military arsenals of countries worldwide, India's security would not in any way be jeopardised. This hypothesis can be further tested in the cases of both Pakistan and China, the two major threats perceived to India's security.

As regards Pakistan, it is a known fact that India is conventionally superior. If anything, Pakistan's quest and acquisition of a nuclear weapons capability has been driven by a desire to offset this superiority. With such a capacity, Pakistan believes it can tie India's hands and curtail its ability to use its conventionally superior force since this would contain the risk of spiralling into a nuclear confrontation. Convinced of this logic, Islamabad believes that its perpetration of cross-border terrorism on Indian soil could continue even as India's retaliatory options are limited. Some merit may be spied in some of these assumptions. And, it is for this reason that the elimination of nuclear weapons makes sense. If nuclear weapons were to be removed from the military strategies of both countries, Indian national security would not be affected given its conventional superiority.

With China, a straight comparison poses a bit of a problem, though it may be said that with some amount of military modernisation, now long overdue, a non-nuclear China would not enjoy any substantial advantage in terms of its conventional strength vis-à-vis a non-nuclear India. Most significantly, no nuclear coercion or blackmail would be possible from China.

Meanwhile, in terms of actual war-fighting, in case the two countries were to confront one another with only conventional weaponry, then a few parameters will have to be kept in mind. First, the fact that the days of total war are virtually over. Any war that the two might fight in the future would most likely be a limited one, in terms of scope, space and time i.e. it will be a limited local war. China, especially, is a proponent of limited war under high tech conditions. In this context, if China were to fight a conventional war with India in order to redeem areas it claims in the northeast and further up in the disputed Himalayan reaches, then the terrain in which the hostilities would occur would provide an inherent advantage to India. Our lines of communication in the area are shorter and more secure. Maintaining a technological edge in Indian weapon systems, besides raising a few more infantry divisions, if required, could further strengthen this advantage. So, the point is that in the absence of a nuclear China, India would withstand a conventional onslaught. Rather, it is the presence of nuclear weapons that tilts the balance in favour of China not only because of its larger nuclear arsenal and deployed tactical nuclear weapons, but also because it has better delivery vehicles.

Therefore, a nuclear free India can more than match its nuclear free neighbours. In fact, while nuclear deterrence is taken to enhance national security for a country faced with a nuclear adversary, it should not be forgotten that there is always a chance, however remote, of the deterrence breaking down. This risk is no less given that the states share contiguous land borders and that too disputed ones, which are prone to everyday frictions. For instance, there can be no denying the fact that during the recent tensions between India and Pakistan in 2002, the presence of nuclear weapons had to be factored into every step that the two countries took, keeping in mind the risk of escalation. The very fact that both nations were armed with nuclear weapons heightened the level of risk of a conflict escalating to a nuclear level. This risk will remain as long as nuclear weapons stay with the countries. This is a frightening prospect because a nuclear war would cause destruction of life on a massive scale and with relative ease.[9] In fact, the destruction so wrought would not be measurable only in terms of numbers of casualties, but would go much further in time and space.[10]

Besides the risk of a conventional confict escalating into a nuclear exchange, the presence of nuclear weapons also carries the threat of a nuclear war brought on by an unintended accident or miscalculation. In fact, modern day communication and sophisticated delivery systems take pride in extremely short lead times between the political decision to use a nuclear weapon and the ability to bring it down on the predesignated target. This carries the inherent risk of a nuclear weapon being launched during a crisis without adequate thought or in the heat of the moment. The danger of unauthorised nuclear launch cannot be dismissed either, since the nascent nuclear command and control structures in India, China and Pakistan cannot be absolutely foolproof.

An environment free of nuclear weapons would, therefore, obviate the risk of nuclear war brought on by an accidental, unauthorised or miscalculated use of nuclear weapons. This can be better explained through an analogy. When driving in chaotic conditions, one has to be responsible for not only one's own driving (which is easy) but also remain constantly alert to driving mistakes that others could make (which is infinitely more difficult) and which could endanger one's own security and existence. By the same logic, if nuclear weapons are available with others in the region, India must not only be able to develop and sustain an effective command and control system for its own weapons, but also hope that others too have foolproof measures in place, But, if there are no nuclear weapons with anyone, it would not only obviate the threat or the actuality of their deliberate use, but also eliminate any possibility of their inadvertent use through an accident or a miscalculation. This is even more crucial given that India, Pakistan and China lie in close vicinity to one another. Radioactive fallout from a nuclear explosion cannot be limited in space and the chance of some adverse effects on the nation that inflicts the misadventure on another cannot be completely ruled out. Given such facts, it seems logical that an NWFW would, as a happy by-product, ensure a safer regional security environment for India.

Nuclear Disarmament and National Development

Sheer economics too prompts India to strive for nuclear disarmament. While it is often argued that nuclear weapons present themselves as low cost security enhancers, it cannot be overlooked that the paraphernalia around them that is required to sustain their operability is not cheap by any measure. Therefore, even if the cost of the nuclear weapon itself may not be very high, the cost of the system as a whole can be exorbitant. The USA, after arguing that nuclear weapons were a cheaper option, ended up spending four trillion dollars on the weapon stockpiles, delivery vehicles and a command, control, communications and intelligence (C^3I) system.[11] Out of this total expenditure, only ten per cent went towards the cost of the weapons themselves, while the rest was spent on delivery systems and C^3I. China claimed that it would have a modest programme but it has spent over $100 billion.[12]

It needs to be realised that nuclear weapons are for defence and deterrence. They do not form a part of war-fighting strategy that must necessarily remain confined to conventional weaponry. Therefore, even in the case where nuclear weapons are available, a high level of conventional capability is still needed to raise the nuclear threshold. Consequently, the cost of nuclear weaponisation is always in addition to the existing expenditures on sustaining and modernising the conventional capabilities of the nation. A logical deduction from the above, therefore, is that the cost of national defence can only go up and not come down with the induction of a nuclear weaponisation programme. The cost of building delivery capabilities and putting a C^3I system in place would pose an additional exorbitant burden.

Of course, one could always argue, and rightly so, that no cost can be high enough when national security is at stake. The nation has to be defended at all costs. But if nuclear weapons could be done away with, without upsetting India's security calculations, then a case can be made for nuclear disarmament actually freeing funds for other developmental activities.

Nuclear Disarmament and International Security

In a unique conceptualisation of nuclear disarmament, India has traditionally viewed nuclear weapons as a universal threat to the entire human race. Therefore, its pursuit of an NWFW has been driven by the need to save mankind from all kinds of potential threats. In fact, Nehru considered the question of disarmament as being more important "than any other problem, internal and external, national or international, because it is a national problem, apart from being an international one."[13] Perhaps no other nation has attached such a connotation to nuclear disarmament, and consequently, none has pursued the goal as fervently either.

The third reason, therefore, why India must strive for the devaluation and elimination of nuclear weapons is because their continued existence carries within it the seed of nuclear proliferation. If those who possess the weapon continue to show a fondness for it, then over time, others too could be expected to feel the need to develop the same. It is not going to be possible to indefinitely profess that a nuclear arsenal is vital to the national security of a few but harmful to global security if others develop or acquire the same. It is in this context that Jonathan Schell has referred to the US arsenal as a "proliferant" and not as a "deterrent",[14] because as long as weapons exist with one, others are going to aspire to possess them.

The acquisition of a nuclear weapon or its technology, in contemporary times, may not prove too difficult given the widespread availability of information and the relatively easy accessibility to fissile material. Of course, the NPT provides some sort of a notional barrier. But, as the examples of Iraq and North Korea have demonstrated, neither the NPT, nor other export control regimes, have been able to restrain those not ready to voluntarily accept limitations on themselves.

While one cannot deny any country the right to defend itself, even if it be through the acquisition of nuclear weapons if it so deems necessary, nuclear proliferation does impose additional strain on international security and stability. Nuclear weapons, by their inherent nature are not weapons of just any other kind. Their possession has to be necessarily accompanied by the development of a fairly sophisticated command and control apparatus so as to minimise, if not totally obviate, the chances of accidental, miscalculated or unauthorised nuclear war. And, it is for this reason that non-proliferation of nuclear weapons becomes a desired objective. But non-proliferation can be sustained only when it is accompanied by the elimination of these weapons.

Another most obvious benefit of universal nuclear disarmament would be a safer world, free from the spectre of deliberate or unintended nuclear war and from the risk of nuclear terrorism. The magnitude of the problem of cross-border terrorism that the country faces requires little elaboration. This could acquire a more sinister *avataar* in case terrorists were to see merit in employing nuclear weapons, however small or crudely designed, into their terrorist ploys. While the threat of nuclear terrorism has not traditionally been taken seriously, it may not be correct to dismiss its possibility, especially in view of the following factors. First, because international terrorism fuelled by religious fundamentalism is on the rise and has become a tool to wage proxy wars. A well-established nexus among international terrorism, organised crime and narco-trafficking ensures a steady supply of resources. Highly motivated and monetarily well-endowed outfits, then, might have little compunction about using a nuclear weapon, or its threat, for certain limited gains. The second factor that raises the probability of nuclear terrorism are the less than adequate storage facilities for fissile material and weapons at the Russian nuclear weapon complexes. Aspersions continue to be cast on the deficiencies in the safe and secure storage of nuclear material. Moreover, given the poor economic conditions of the nuclear workers, the chances of a deal being struck for purely monetary reasons cannot be ignored. Also, the accessibility to nuclear technology has increased with several workers in what was once a huge nuclear weapons complex now becoming unemployed or badly underpaid. In a situation like this, if terrorists so desire, obtaining nuclear material, components, technology or even a weapon may not be an impossible task.

In the case of India, this problem acquires a more serious dimension given that the actual custody of nuclear weapons in Pakistan is believed to be with the military and this is known to overtly support the *jehadi* terrorists. Therefore, in order to remove, or at least, substantially minimise the spectre of nuclear terrorism, nuclear disarmament presents itself as an attractive option.

Apart from this, a nuclear weapon-free world would bring India the additional advantage of bringing about the demise of an iniquitous world order that the country has long been fighting against. India has always campaigned against the discriminatory non-proliferation order that the NWS have sought to impose on the international community through the NPT and its safeguards regime. The distribution of unequal rights and obligations under these mechanisms have, therefore, never been accepted by India. Universal nuclear disarmament would remove nuclear apartheid.

Nuclear Disarmament: A Legal Obligation

Yet another reason why India must not abandon its pursuit of nuclear disarmament is because it is a legal obligation upon its shoulders as a state with nuclear weapons to strive for the elusive goal. Though the NPT does not grant it recognition as an NWS, the fact of the matter is that India, as also Pakistan, do possess nuclear weapons capability and, therefore, are nuclear weapon states. Striving for the goal of nuclear disarmament then becomes a legal responsibility given the 1996 ruling of the ICJ which states, "There exists an obligation to pursue in good faith and bring to a conclusion negotiations leading to nuclear disarmament in all aspects, under strict international control."[15] Therefore, India must respect the decision of the ICJ to the best of its abilities and continue to perform its role as a law-abiding member of the international comity of nations.

Nuclear Disarmament: An Environmental Imperative

Overt nuclear weapons capability implies environmental costs arising from nuclear tests, possibility of accidents, decommissioning of old nuclear warheads, clean-up of nuclear production facilities and the impact of weapons, if ever used. Even if regular nuclear testing is not done, thereby obviating the hazards of radioactive contamination, accidents could still occur during the various activities involved in the production of nuclear warheads. Subsequently, the decommissioning of older nuclear warheads and the cleaning up of contaminated nuclear facilities poses additional environmental challenges. Above all, if ever the weapons happen to be used, the effects on human health and the environment cannot even be quantified. For South Asia, the consequences would be even more immediate and inescapable given the geographical contiguity of the nuclear adversaries. The country launching the first nuclear strike would not be able to escape some amount of ruin from the radioactive fallout, even in the unlikely event that the other side does not mount a nuclear retaliation.

Conclusion

From the above arguments, it can be deduced that nuclear disarmament would not be detrimental to India's core interests of national security. Rather, it would bring several other advantages—economic, environmental and moral. In present day India, when contemporary emphasis is on national strength and when only the moral justification for nuclear disarmament is not enough by itself, this chapter has sought to highlight that even on a realistic level, nuclear disarmament remains the need of the hour. In the past, India has used this plank to rally support from among the ranks of the non-aligned nations. Disillusionment in these nations on the real motives and intentions of the NWS is rampant, now more than ever. India could still stand to garner their support if it were to make a renewed commitment and exhibit a visionary approach in a collective march towards universal nuclear disarmament.

Having established that the attainment of nuclear disarmament would be in India's national interest, it becomes imperative that the country should pursue the objective in a proactive manner. The attainment of a secure and stable nuclear weapon-free world would naturally call for a large number of measures. These would have to range from fostering greater mutual trust in inter-state relations, a greater transparency in defence matters, more encompassing arms control efforts that specify deeper cuts to a stricter implementation and compliance with all treaties related to non-proliferation and disarmament. However, before embarking upon the tangible steps, it would be crucial to target the mindsets and beliefs of those against nuclear weapons abolition. In fact, this would have to be the essential starting point because unless we begin to think differently about nuclear weapons, we cannot be expected to act differently with them. Is it not said, "Since wars begin in the minds of men, it is in the minds of men that defences of peace must be constructed."[16]

In the ultimate analysis, it may be reiterated that nuclear disarmament is not going to be an easy or quick process. At the same time, it may not be assumed that it is impractical or impossible. The prospects might not be bright right now but things could begin to look up if international security is enhanced through more CBMs; if an effective and plausible verification mechanism is evolved to enforce disarmament and ensure compliance; if the risks of nuclear weapons possession outweigh the benefits they are supposed to render; and if enough moral pressure is mounted by the saner voices. All these conditions are not unattainable.

In an NWFW, India can be regionally more secure and globally better placed. It must, therefore, remain on the look-out for every possible window of opportunity. One such possibility may be discerned in the fact that the NNWS within the NPT are becoming increasingly disenchanted with the treaty and the kind of non-proliferation it practises. Consequently, a number of important initiatives have been taken by the NNWS as well as NGOs to outline some proposals for an eventual elimination of nuclear weapons.

For instance, in 1998, after the inability of the second NPT PrepCom to make any progress on substantive issues, a grouping of eight countries called the New Agenda Coalition (NAC) was brought into being. Brazil, Egypt, Ireland, Mexico, New Zealand, South Africa and Sweden decided to pool their resources, ideas and capabilities to present workable solutions to issues deadlocking negotiations. In fact, in the same year, it had even presented a detailed road map to achieving an NWFW to the UN. The UNGA then adopted the resolution with 114 in favour, 18 against and 38 abstentions.[17]

Pressing on with its efforts, at the 2000 NPT RevCon, the NAC presented a working paper, forcefully projecting their case for nuclear disarmament by:

(a) calling upon the five NWS to make an unequivocal commitment to give up their nuclear arsenals;[18] and

(b) suggesting measures such as the formulation of nuclear policies and postures so as to preclude the use of nuclear weapons; to de-alert and remove nuclear warheads from delivery vehicles; and, to reduce tactical nuclear weapons and proceed to their elimination as an integral part of nuclear arms reductions. The proposed set of measures also includes the need for a greater transparency with regard to nuclear arsenals and fissile material inventories.

Of course, such proposals have not fructified yet. But they do help generate positive impulses that can be worked upon by, or with, others. It is in this regard that India can let it be known that it shall lend open support to practical proposals and initiatives that would genuinely lead the world towards nuclear abolition. In the ultimate analysis, every voice in favour of nuclear disarmament will count and given India's gains in a nuclear weapon-free world, it must keep its voice loud and clear.

Notes

1. The NPT recognises only those countries as nuclear weapon states that had tested their nuclear devices before January 1, 1967.
2. A confidential note written at Wardha on August 25, 1940, and as quoted in "Non-Violence and the State," *India and Disarmament: An Anthology* (New Delhi: Ministry of External Affairs, Government of India, 1988), p. 11.
3. J.P. Morray, *From Yalta to Disarmament: Cold War Debate* (New York: MR Press, 1961), p. 275.
4. Rajendra Prasad, "The Case for Unilateral Disarmament," in Ibid., p. 139.

5. "Delhi Declaration," in Ibid., p. 82.
6. The US literally established this in November 1997, when the Presidential Decision Directive 60 affirmed US reliance on nuclear arms as a cornerstone of its national security.
7. "Essentials of Post-Cold War Deterrence," in Hans M. Kristensen, *Nuclear Futures: Proliferation of Weapons of Mass Destruction and US Nuclear Strategy,* BASIC Research Report 98.2, March 1998.
8. The Indian polity found itself divided over whether China's test should be responded to with an Indian test soon thereafter or not.
9. It can be argued that modern day conventional weapons can cause a similar extent of destruction. But the difference lies in the fact that to do so would require a more intensive and extensive use of air power. Nuclear weapons, meanwhile, offer the temptation of easy use and, therefore, also invoke the fear of irresponsible use.
10. For a detailed description of the kind and scale of destruction and damage a nuclear attack could cause, see Jonathan Schell, *The Fate of Earth* (London: Picador, 1982).
11. "Nuclear Arms, the Budget and the Economy," Draft prepared for a discussion in Movement in India for Nuclear Disarmament (MIND).
12. Praful Bidwai and Achin Vanaik, "India's Nuclear Daze: The Domestic Politics of Nuclearisation," TNI and Institute for Policy Studies, Washington, August 1998.
13. India, Lok Sabha, *Debates,* November 23, 1960, pp. 1937-40.
14. Jonathan Schell, "The Folly of Arms Control," *Foreign Affairs,* Autumn 2000.
15. Legality of the Threat or Use of Nuclear Weapons, Advisory Opinion, Communique No. 96/23, July 8, 1996, International Court of Justice, The Hague.
16. The quote is borrowed from the UNESCO Constitution. For a more detailed study on this aspect of the power of the mind over matter, see Manpreet Sethi, "Practical Steps to Abolish Nuclear Weapons: Addressing the Belief Systems," paper originally presented at the Pugwash workshop on the Abolition of Nuclear Weapons held at New Delhi, on October 11-13, 1999, and published in *Strategic Analysis,* vol. XXIII, no. 9, December 1999.
17. Of the 16 NATO states, only the US, UK, France and Turkey voted against the resolution. The others abstained. Canada was even prepared to vote in favour of the resolution if another NATO country did likewise. But Germany, Netherlands, Norway, Belgium and Japan succumbed to domestic pressure and abstained. Also see William Epstein, "Voting by Abstention", *Bulletin of Atomic Scientists,* March-April 1999, p. 6, and Rebecca Johnson, "Troubled Treaties: Is the NPT Tottering?" http://www.basicint.org. A series of reports

on the PrepComs and RevCon were compiled by Rebecca Johnson and are available on this website.

18. Such a call was made at the 2000 NPT RevCon. However, it remained as ambiguous as similar calls made in the past, since it projected no time-frames in which the NWS would implement measures towards nuclear disarmament. Rather, the countries, cleverly couched and premised all further progress in this direction on the continuance of the conditions of "strategic stability"—a phrase open to any number of interpretations.